Lecture Notes in Control and Information Sciences

Edited by M. Thoma

Vol. 22: Optimization Techniques
Proceedings of the 9th IFIP Conference on
Optimization Techniques,
Warsaw, September 4–8, 1979
Part 1
Edited by K. Iracki, K. Malanowski, S. Walukiewicz
XVI, 569 pages. 1980

Vol. 23: Optimization Techniques
Proceedings of the 9th IFIP Conference on
Optimization Techniques,
Warsaw, September 4-8, 1979
Part 2
Edited by K. Iracki, K. Malanowski, S. Walukiewicz
XV, 621 pages. 1980

Vol. 24: Methods and Applications
in Adaptive Control
Proceedings of an International Symposium
Bochum, 1980
Edited by H. Unbehauen
VI, 309 pages. 1980

Vol. 25: Stochastic Differential Systems –
Filtering and Control
Proceedings of the IFIP-WG7/1 Working Conference
Vilnius, Lithuania, USSR, Aug. 28 – Sept. 2, 1978
Edited by B. Grigelionis
X, 362 pages. 1980

Vol. 26: D. L. Iglehart, G. S. Shedler
Regenerative Simulation of Response
Times in Networks of Queues
XII, 204 pages. 1980

Vol. 27: D. H. Jacobson, D. H. Martin, M. Pachter, T. Geveci
Extensions of Linear-Quadratic Control Theory
XI, 288 pages. 1980

Vol. 28: Analysis and Optimization of Systems
Proceedings of the Fourth International
Conference on Analysis and Optimization of Systems
Versailles, December 16–19, 1980
Edited by A. Bensoussan and J. L. Lions
XIV, 999 pages. 1980

Vol. 29: M. Vidyasagar,
Input-Output Analysis of Large-Scale
Interconnected Systems –
Decomposition, Well-Posedness and Stability
VI, 221 pages. 1981

Vol. 30: Optimization and Optimal Control
Proceedings of a Conference Held at
Oberwolfach, March 16–22, 1980
Edited by A. Auslender, W. Oettli, and J. Stoer
VIII, 254 pages. 1981

Vol. 31: Berc Rustem
Projection Methods in Constrained
Optimisation and Applications
to Optimal Policy Decisions
XV, 315 pages. 1981

Vol. 32: Tsuyoshi Matsuo,
Realization Theory of
Continuous-Time Dynamical Systems
VI, 329 pages, 1981

Vol. 33: Peter Dransfield
Hydraulic Control Systems –
Design and Analysis of Their Dynamics
VII, 227 pages, 1981

Vol. 34: H.W. Knobloch
Higher Order Necessary Conditions
in Optimal Control Theory
V, 173 pages, 1981

Vol. 35: Global Modelling
Proceedings of the IFIP-WG 7/1 Working
Conference Dubrovnik, Yugoslavia,
Sept. 1–5, 1980
Edited by S. Krčevinac
VIII, 232 pages, 1981

Vol. 36: Stochastic Differential Systems
Proceedings of the 3rd IFIP-WG 7/1
Working Conference
Visegrád, Hungary, Sept. 15–20, 1980
Edited by M. Arató, D. Vermes, A.V. Balakrishnan
VI, 238 pages, 1981

Vol. 37: Rüdiger Schmidt
Advances in Nonlinear
Parameter Optimization
VI, 159 pages, 1982

Vol. 38: System Modeling and Optimization
Proceedings of the 10th IFIP Conference
New York City, USA, Aug. 31 – Sept. 4, 1981
Edited by R.F. Drenick and F. Kozin
XI, 894 pages. 1982

Vol. 39: Feedback Control of
Linear and Nonlinear Systems
Proceedings of the Joint Workshop
on Feedback and Synthesis of
Linear and Nonlinear Systems
Bielefeld/Rom
XIII, 284 pages. 1982

Vol. 40: Y.S. Hung, A.G.J. MacFarlane
Multivariable Feedback:
A Quasi-Classical Approach
X, 182 pages. 1982

Vol. 41: M. Gössel
Nonlinear Time-Discrete Systems –
A General Approach by
Nonlinear Superposition
VIII, 112 pages. 1982

Vol. 42: Advances in Filtering and
Optimal Stochastic Control
Proceedings of the IFIP-WG 7/1
Working Conference
Cocoyoc, Mexico, February 1–6, 1982
VIII, 391 pages. 1982

For information about Vols. 1–21 please contact your bookseller or Springer-Verlag.

Lecture Notes in Control and Information Sciences

Edited by M. Thoma

DFVLR

70

Uncertainty and Control

Proceedings of an International Seminar Organized by Deutsche Forschungs- und Versuchsanstalt für Luft- und Raumfahrt (DFVLR) Bonn, Germany, May, 1985

Edited by J. Ackermann

Springer-Verlag Berlin Heidelberg GmbH

Editor
Dr.-Ing. Jürgen Ackermann
Institut für Dynamik der Flugsysteme
Deutsche Forschungs- und Versuchsanstalt
für Luft- und Raumfahrt (DFVLR)
D-8031 Wessling/Oberpfaffenhofen
Germany

ISBN 978-3-540-15533-1 ISBN 978-3-540-39448-8 (eBook)
DOI 10.1007/978-3-540-39448-8

Originally published by Springer-Verlag Berlin Heidelberg New York in 1985.
Offsetprinting: Mercedes-Druck, Berlin
2061/3020-543210

P R E F A C E

The German Aerospace Research Establishment - Deutsche Forschungs- und Versuchsanstalt für Luft- und Raumfahrt e.V. (DFVLR) - has initiated a new series of seminars concerning fundamental problems in applied engineering sciences. These seminars will be devoted to interdisciplinary topics related to the vast variety of DFVLR activities in the fields of fluid mechanics, flight mechanics, guidance and control, materials and structures, non-nuclear energetics, communication technology, and remote sensing.

The purpose of the series is to bring modern ideas and techniques to the attention of the DFVLR in order to stimulate internal activities, and to promulgate DFVLR achievements within the international scientific/technical community. To this end, prominent speakers are invited to join in a series of lectures and discussions on topics of mutual interest.

After the first seminar in 1984 on "Nonlinear Dynamics in Transcritical Flows" this second seminar deals with "Uncertainty and Control" in general dynamic systems.

The dividing line between a "system" and its environment is not well defined. Nevertheless, the line has to be drawn in physical/mathematical modelling. As a result, unmodelled events in the environment lead to uncertainties in the structure and parameters of the system or introduce noise. A control system tries to enforce a specified dynamic behavior in spite of this uncertainty. Feedback of sensor signals in robust, adaptive and intelligent control systems can reduce the effect of uncertainties.

In this seminar, the state of the art of the theory for the design of such control systems will be surveyed and the relation with applied engineering problems and solutions will be discussed.

Bonn, 7. Mai 1985

Prof. Dr. H.L. Jordan
Chairman of the
Board of Directors DFVLR

CONTENTS

UNCERTAINTY AND CONTROL
-SOME ACTIVITIES AT DFVLR

compiled by

Georg Grübel
DFVLR-Institute for Flight Systems Dynamics
D-8031 Oberpfaffenhofen

Summary

Some activities at the DFVLR which deal with system modelling and performance evaluation under uncertainty, as well as feedback control applications, are reported. The examples are mainly drawn from activities of the research department 'Flight Mechanics / Guidance and Control' where active control of aircraft and helicopters plays a central role. Uncertainty is related to system parameters, disturbances, operational state and operator behaviour. In this respect, DFVLR activities in applied nonlinear parameter identification, on-line wind measurement and prediction, stochastic simulation, and sensor diagnosis via analytic redundancy are briefly described. Applications of feedback control concern model-following control for inflight simulation, robust stabilization of high-performance aircraft, aircraft flutter stability augmentation via active mode decoupling, and active damping of mechanical light-weight structures based on finite-element modelling.

The DFVLR -the German Aerospace Research Establishment- is the largest research establishment dealing with engineering sciences in the Federal Republic of Germany. Beyond its main activities in aerospace, the DFVLR is applying its know-how to other fields of engineering, where related technical problems have to be solved. These fields include energetics and propulsion, environment engineering, advanced ground transportation and robotics. The DFVLR has the most up to date test facilities and technical installations at its disposal.

In the following, some activities at the DFVLR are compiled which relate to the topic "Uncertainty and Control" as it is understood in this seminar. The list of these activities is by no means complete. The examples are mainly drawn from activities of the research department 'Flight Mechanics / Guidance and Control' where active control of aircraft and helicopters plays a central role.

VARIOUS ASPECTS OF UNCERTAINTY

We consider the main information and control structure of a piloted aircraft as depicted in Figure 1.

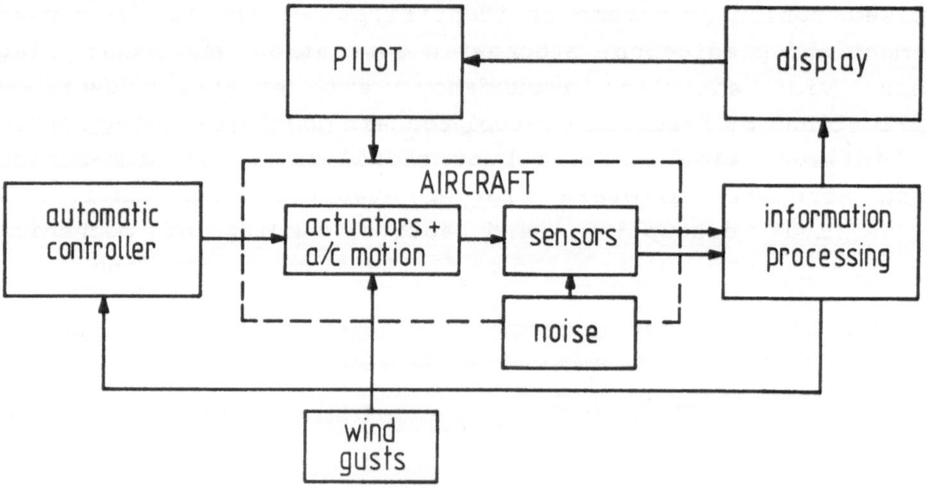

Figure 1: Information and control structure of a piloted aircraft.

Uncertainty in this pilot-aircraft control loop may be related to

Uncertainty in system parameters
Uncertainty in system parameters is essentially related to the equations of motion and the dynamic responses of actuators (and sensors). Using the basic laws of physics and engineering, the structure of the equations is assumed to be known but the parameters have to be "identified" and they are known only within tolerances.

Uncertainty due to disturbances
Uncertainty due to disturbances may be divided into two categories: unknown effects of the environment such as wind gusts ("external" disturbances) and disturbances generated by the technical system itself such as sensor noise ("internal" disturbances).

Uncertainty in the operational state
Uncertainty in the operational state may be considered on two levels: sensor failures in automatic control loops and information-display failures in man-machine decision/control loops.

Uncertainty in operator behaviour
Uncertainty in operator behaviour may be attributed to the questions: How does a human operator behave as a decision-maker to manage slow "situation dynamics" and how does the operator behave as an in-the-loop controller to deal with fast motion dynamics?

These four aspects of uncertainty in a pilot-aircraft control loop are typical for general man-machine systems. We take them as guideline to survey some DFVLR-activities which aim at reducing such kind of uncertainty via proper modelling, performance prediction and system layout. Such activities are: (nonlinear) parameter identification to determine system parameters as accurately as possible, on-line wind measurement and prediction using Kalman filtering, stochastic simulation to predict nonlinear system performance under internal and external disturbances, and sensor diagnosis via analytic redundancy.

PARAMETER IDENTIFICATION

The increasingly demanding requirements imposed on aircraft performance and flight characteristics call for an extended use of automatic control by implementing active control systems. The use of active control technology however requires a detailed understanding of the influence of the anticipated external disturbances, aerodynamic characteristics and control system responses and requires considerable advances in the ability to describe and model such phenomena.

While the estimation of *static* flight mechanics coefficients normally involves no difficulties, problems may arise in the experimental estimation of *dynamic* stability parameters. Even more difficult is the identification of those effects which are related to high frequency motions as in the case of flight in turbulence. These effects, called the *instationary aerodynamic transient effects*, can play an important role in the design of an active control system like a gust alleviation system [1].

The field of aircraft *parameter identification* and dynamic flight testing has by now emerged as an essential tool for the evaluation of prototype aircraft. Applications of this technique are [12]

- the evaluation of aircraft performance characteristics such as polar drag curve or steady state climb performance in non-steady flight conditions,

- the identification of nonlinear mathematical models of aerodynamic forces and moments, and of primary control force characteristics for aircraft flight simulation, as well as,

- the estimation of linear stability and control derivatives for the numerical assessment of flying qualities and for the design of linear control systems.

Parameter identification is based on the following idea (cf. Figure 2): Assume that the dynamical input-output behaviour of a physical system be mathematically described by (nonlinear) equations (1), (2) with yet undetermined parameters β. By exciting the physical system by appropriate input signals u(t), measuring the physical system response as $z(t_k) = y(t_k) + \mu(t_k)$ (usually with some uncertainty due to measurement noise μ) and comparing this measured response with the corresponding "response" of the mathematical model, one can try to fit the physical and mathematical responses by suitably adjusting the free parameters β.

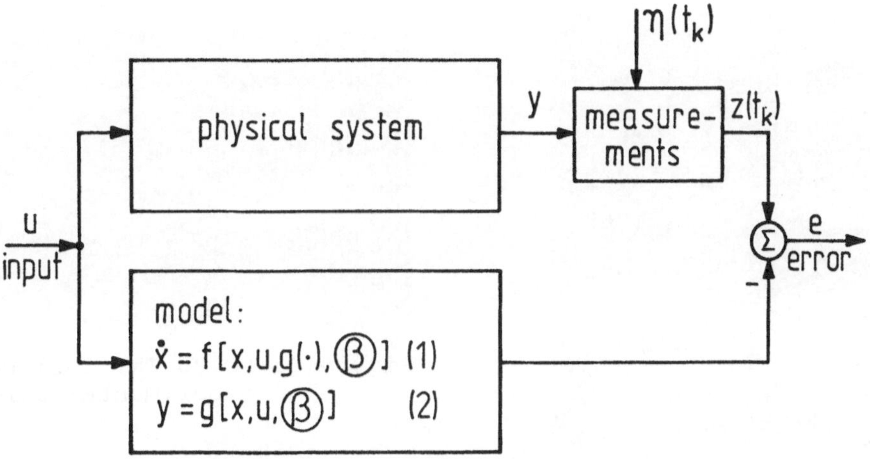

Figure 2: Basic scheme for parameter identification.

Sucessful parameter identification requires expertise in the following fields:

- *experimental set up*
- *input signal design*
- *mathematical identification technique*
- *computational efficiency.*
- *model verification*

Experimental set up

As an experimental set up for aircraft and helicopter parameter identification essentially two approaches are pursued at the DFVLR: sub-scale dynamic wind tunnel investigations (Figure 3) and full-scale free flight tests (Figure 4) [1]. In both cases expert experimental planning and adequate instrumentation is required which covers the entire data aquisition process including measurement noise and sensor dynamics.

Figure 4: DO 28 TNT full-scale experimental aircraft

Figure 3: DO 28 TNT 1/8-scale model in DFVLR wind tunnel.

Input signal design

The input signals must be designed such that all relevant system response modes are excited. They have to be "optimized" to achieve accurate identifications. The shape of an input signal which is both simple to apply and very efficient for parameter identification of the longitudinal motion of aircraft is the DFVLR 3-2-1-1 signal for elevator deflections [11] (Figure 5). Time-scaling of this signal is chosen such that the maximum value of the input power spectrum is located near the short period oscillation frequency (approx. 1 Hz) (Figure 6).

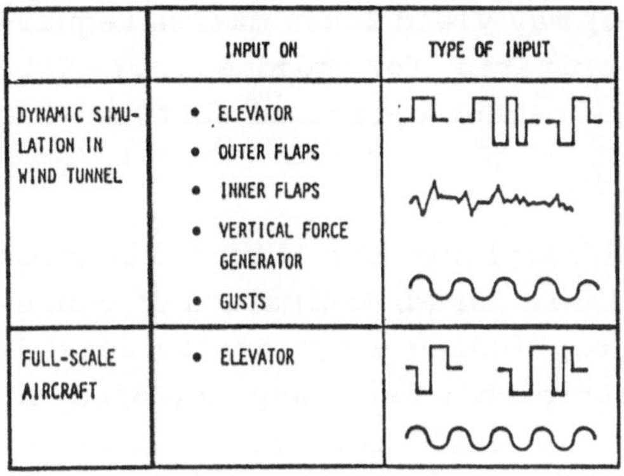
INPUT ON
TYPE OF INPUT
DYNAMIC SIMULATION IN WIND TUNNEL
• ELEVATOR
• OUTER FLAPS
• INNER FLAPS
• VERTICAL FORCE GENERATOR
• GUSTS
FULL-SCALE AIRCRAFT
• ELEVATOR

Figure 5: Input signals for aircraft parameter identification : 3-2-1-1 signal for elevator deflection [1].

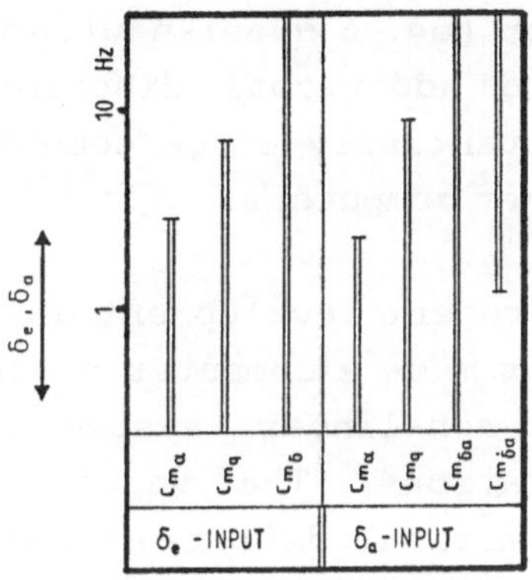

Figure 6: Frequency range of input signals for elevator and aileron necessary for aircraft parameter identification [1].

Mathematical identification technique

The standard mathematical identification technique used at the DFVLR is maximum-likelihood estimation (ML) (cf. Figure 7).

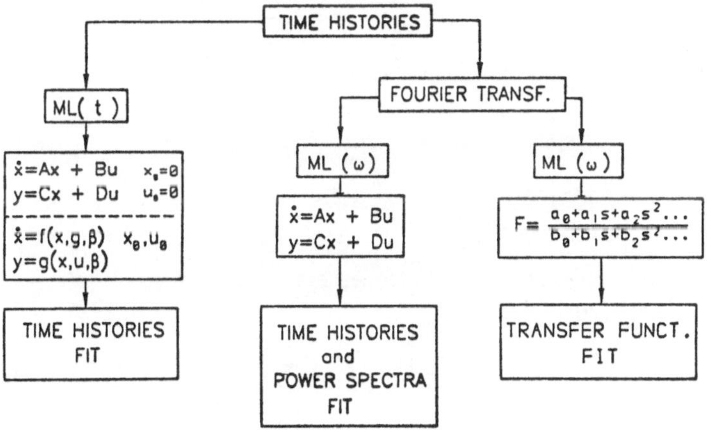

Figure 7: Parameter identification techniques used at DFVLR (ML = Maximum Likelihood estimation).

Features of the maximum-likelihood identification technique are: It can deal with nonlinear models with noise and the data compatibility check, which is always necessary, can be handled as an integral part. On the other hand maximization by a local optimizer (eg. a quasi-newton technique) may yield local maxima requiring additional discrimination criteria for goodness of fit. Potentially large computing times demand optimized software and fast computers.

A recent development at the DFVLR-Institute for Flight Mechanics has made a computer program available for ML estimation of general non-linear systems without complicated programmatic considerations. The sensitivity coefficients are approximated by numerical differences formed through small perturbations of the parameters. Through the selective application of two different integration algorithms, systems with discontinuities can be easily handled. Practical identification experience has shown this technique to provide accurate results for both linear and non-linear systems for data obtained from both wind-tunnels and flight tests [13,17].

Computational efficiency

As far as computing time is concerned, a particularly ambitions task is the parameter identification of helicopters (Figure 8).

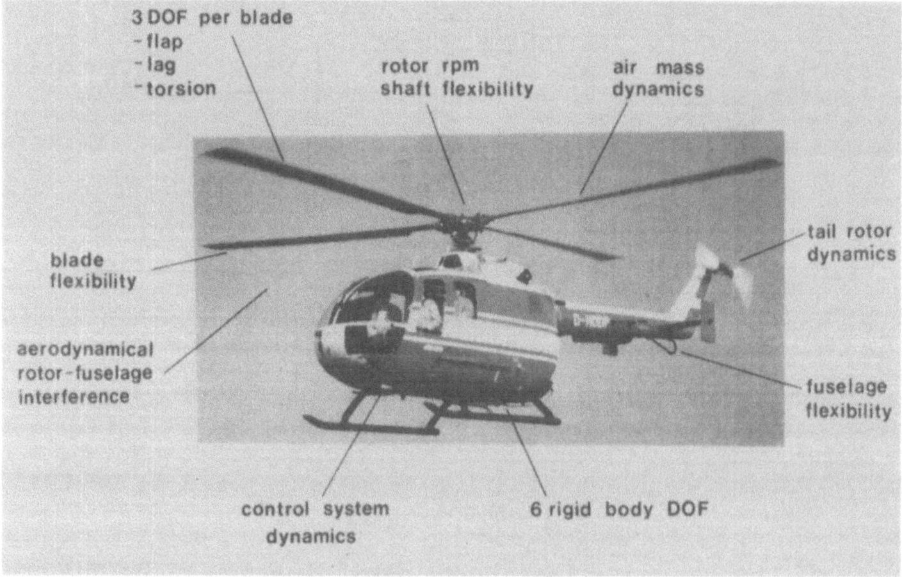

Figure 8: Helicopter dynamic degrees of freedom.

Typically one may have 8 nonlinear state equations and 11 observation equations with 95 parameters to be identified, including parameters for a data compatibility check. With optimized software one complete identification using 2000 data points requires at the DFVLR an overall CPU-time of less than 300 seconds on a CRAY-1/S (80 MIPS) computer [18].

Model verification

Figure 9 shows model verification results if linear or nonlinear parameter identification is used. Needless to say, model verification has to be performed using different data to that used for identification. As far as the parameters are concerned, confidence intervals for the parameter ensembles describing the entire model are of interest rather than confidence intervals for the individual parameters themselves. Such confidence intervals can be determined using Monte Carlo techniques.

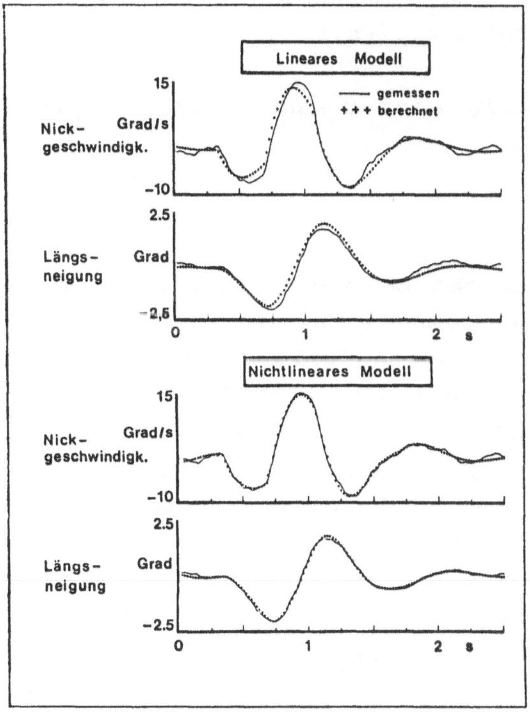

Figure 9:
Model verification results - parameter identification with linear and nonlinear models of DO 28 TNT for pitch rate (=Nickgeschw.) and angle of attack (=Längsneigung) [19].

"EXTERNAL" DISTURBANCES: TURBULENCE

Referring to Figure 1 one may distinguish between two kinds of disturbances: "external" disturbances which are unwanted effects of the environment such as turbulence (wind, gusts) and "internal" disturbances which are generated by the technical system itself such as sensor noise. First let us ask how to deal with turbulence.

The effect of turbulence on flying qualities is of basic importance. Therefore any evaluation of flight performance using simulation should at least include cases of "Moderate" turbulence.

There are various ways to deal with *turbulence uncertainty:*

- *agree upon a "standard" mathematical atmospheric turbulence model and simulate it on a computer,*
- *rely on a physical turbulence simulator for ground based (e.g. wind tunnel) experimental analysis,*
- *use ground-based or on-board wind measurements and predictions via Kalman filtering.*

Standard turbulence models

Atmospheric disturbance models can be separated according to their degree of determinism or randomness. While characteristics such as mean wind shear are normally handled on a deterministic basis, turbulence is usually modelled as a randomly occurring phenomenon. Nevertheless, wind velocity or wind shear can be just as well described in strictly probabilistic terms, and turbulence, conversely, can be described in wholly deterministic terms (as with gusts composed of summed sinusoids). In addition, random and deterministic models are often combined to suit the needs of a particular application. Appropriate partition of model determinism versus randomness figures greatly in the success of any given application.

Deterministic features are usually quantified directly using analytic functions or tables (e.g., mean wind speed and direction or wind shear gradients with respect to time or space). Random components, on the other hand, involve random variable sources having their own particular statistical properties of probability distribution and correlation.

Practical application requires a rational approach to the tradeoff between engineering convenience and physical correctness in disturbance models. For example, the well-known von Karman turbulence form yields more correct spectral characteristics, but it is not so easily realized computationally as the more approximate Dryden form (Figure 10). Practical problems in general include: digital implementation of continuous spectral forms, correct scaling of random noise sources, implementation of gust gradients, gust time derivatives, and gust transport lags.

Notes:

- **Area under curves for a given frequency band corresponds to spectral power in that frequency band**
- **Total aera under either curve is .4343 if unit on abscissa is taken as one decade**
- **Spatial frequency, Ω related to temporal frequency, ω by airspeed, V or $\Omega = \omega/V$**
- **$\Phi(\Omega)$ is power spectral density**
- **L_u is scale length**

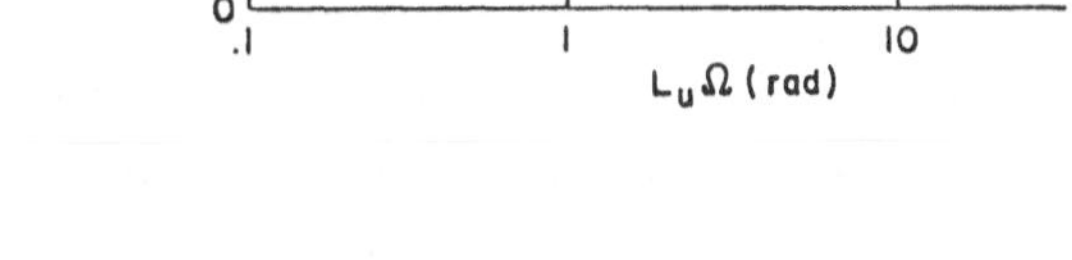

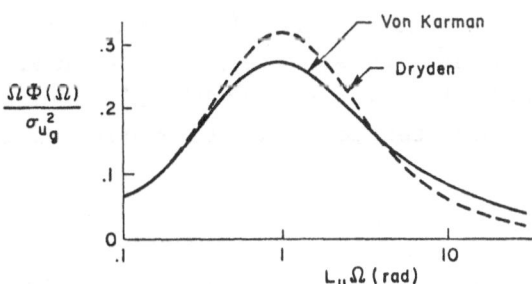

Figure 10:
Power spectral density for Dryden and v. Karman forms of horizontal gust, multiplied by frequency [from STI Proposed MIL Handbook, Dec. 1981].

To develop, qualify and find agreement on "standard" aircraft turbulence models requires international cooperation such as the efforts for the new US MIL Specifications F-8785 C. Here, the DFVLR also plays an active role.

On-line wind measurement and prediction

Sufficiently accurate knowledge of the wind in magnitude and direction as a function of airspace position and time is an essential part not only for navigation but also for fuel efficiency and efficient planning and control of air traffic. It is also a prerequisite for four-dimensional guidance. The calculation of the three dimensional flight path command subject to time-of-arrival constraints (fourth dimension) requires wind prediction along the complete flight path the aircraft is going to track. Since the flight path is to be calculated and updated, on-line wind information has to be available at any time for the respective three dimensional airspace.

Wind data, currently available through meteorological service broadcasts, is not sufficient when used as a single source. This data suffers from small measurement rates with respect to location and time. An actual wind situation can deviate significantly from that announced by the meteorological service. Therefore, complementary or even quite different techniques emerge for wind estimation, in particular using modern aircraft sensor and computer equipment.

The DFVLR-Institute of Flight Guidance has developed such airborne wind estimation techniques. These were implemented in the automatic digital flight control system of the DFVLR's test aircraft HFB 320 and were flight tested. A brief report of this activity is drawn from [25]:

Wind measurements are noisy, partly caused by measurement errors. This requires filtering of the measured wind data. A wind model serves as part of the filtering to structure the mean wind relationships. The parameters of this model are the crucial quantities to be determined as well as the errors of these model parameters. The best suited filter for this purpose is the Kalman filter which processes the difference between the low frequency wind model output and the high frequency wind measurement signal

(Figure 11). The Kalman filter also provides an estimate of the errors of the wind model.

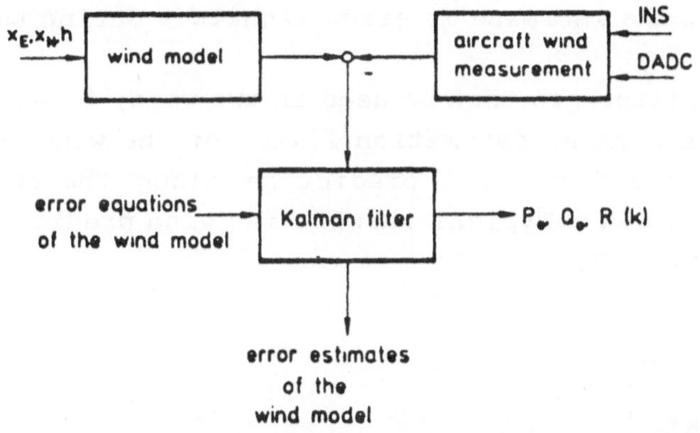

Figure 11: Principle of Kalman Filtering for wind measurements [25].

In order to estimate the wind model errors with sufficient accuracy, additional information for the filtering is needed about those errors which are not described by the error equations. These unmodelled errors are accounted for by the system noise matrix. Also the statistics of the measurement errors, like those of the Inertial Navigation System and air data sensor errors, have to be taken into account. This information is

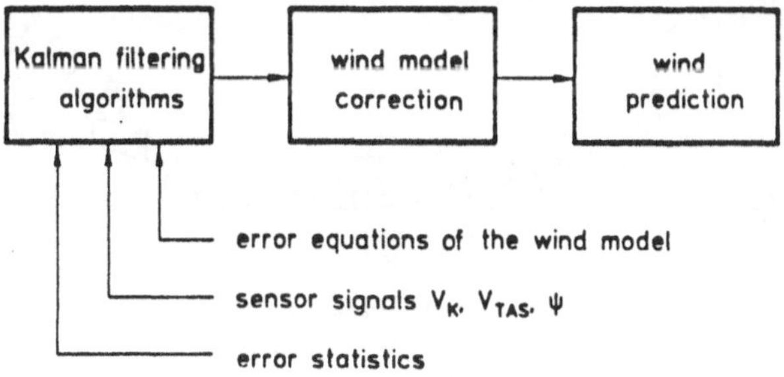

Figure 12: Principle of wind prediction through extrapolation using Kalman Filtering [25].

derived from statistical data of wind profiles and known statistical error models of the sensors. The measurement noise was defined as a discrete function of the bank angle in order to take into account the increase in error magnitude during turns.

The Kalman filter can then be used in two ways, first as smoothing filter and second as estimation filter of the wind model parameters with regard to wind prediction along the 4D-flight path ahead (Figure 12). Typical results for wind prediction are shown in Figure 13a, 13b:

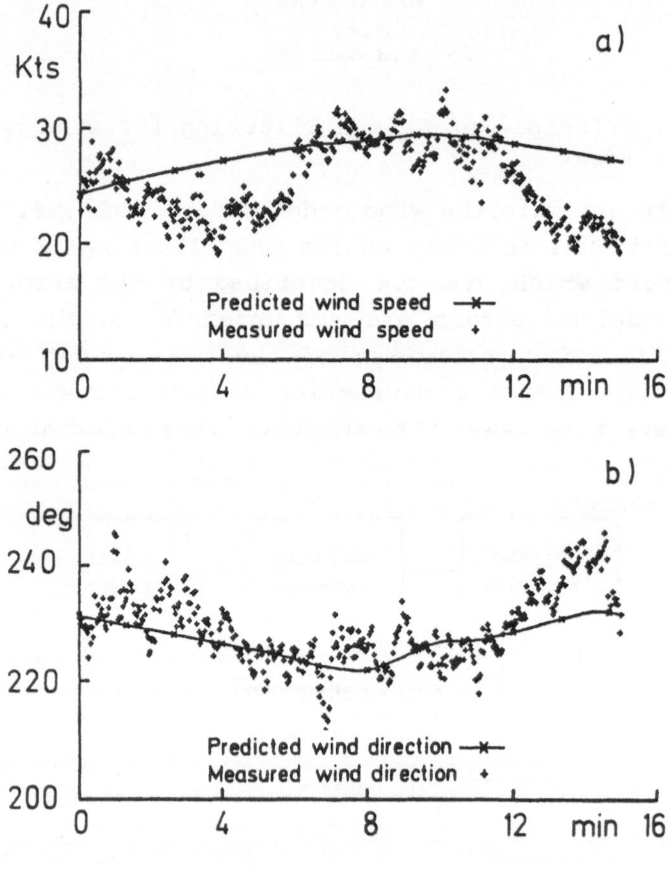

Figure 13: Example for wind prediction through extrapolation using Kalman Filtering [25], a) wind speed, b) wind direction.

"INTERNAL" DISTURBANCES: MEASUREMENT NOISE

Measurements are always subject to some kind of uncertainty. This uncertainty is due to thermal noise of electronic sensors, glint of reflecting surfaces or other physical disturbances associated with the measurement process. If the process, of which measurements have to be taken, and the sensor dynamics can be modelled by linear equations and if the disturbances are assumed to be stochastic with gaussian distribution, then linear Kalman filtering provides optimal estimates of the measured quantities. On-line wind estimation as described above is an application example.

In some systems measurement noise is the main disturbance affecting system performance. Systems of this kind are radar guided homing missiles where range dependent radar glint [28] is the main source of random uncertainty. Furthermore these systems are generically nonlinear because of their kinematic relations and because actuator command limiters (Figure 15) are always present.

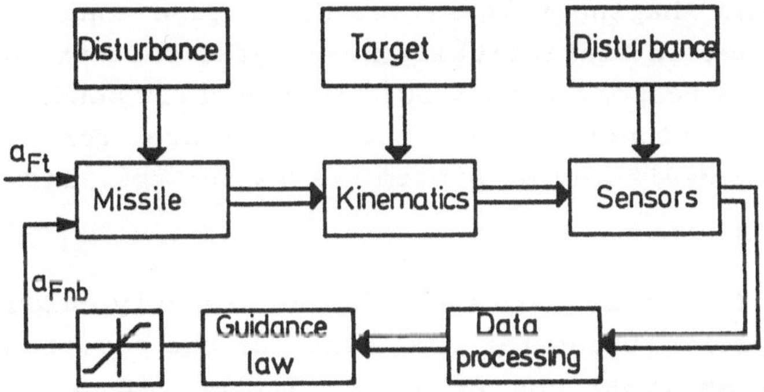

Figure 15: Nonlinear missile guidance loop with sensor disturbances (e.g. radar glint) and missile disturbances (e.g. random initial conditions)[30].

Optimal estimation in this case would require nonlinear filtering. However nonlinear filtering theory has not yet reached the stage where it can be applied in real time to complex systems such as missile guidance loops. How can one proceed?

One possibility is to neglect nonlinear effects and apply Taylor linearization along a prespecified trajectory in order to meet the assumptions of (linear) Kalman filtering theory [32]. Alternatively, one can use "extended" Kalman filtering where a more sophisticated linearization about the estimated variables is used. However, in both cases there is no guarantee that the filter converges in the nonlinear environment. Therefore the actual nonlinear performance has to be assessed by stochastic simulation, taking into account the nonlinear effects neglected during design.

STOCHASTIC SIMULATION

Traditionally, nonlinear stochastic systems are simulated by applying *Monte Carlo techniques*. Let the nonlinear system be described by

$$\dot{x}(t) = f[x(t),t] + B(t)w(t)$$

driven by a random process w(t) and with random initial conditions. Using shaping filters, we can assume that w(t) is white noise. To obtain statistically meaningful results, a great number of sample responses (say 50) are generally needed, leading to a high computational burden. The results are, for example, mean and standard deviation of system states for specific instants of time.

Covariance techniques are methods for directly calculating analytical approximations for the means and the covariances of system states as functions of time. Instead of calculating many sample responses, the state equation above is linearized around its mean value and the resulting (nonlinear) system of equations for the means and the covariances is solved numerically only once. The computing time may thus be reduced. For linearization two methods can be applied: Taylor series and statistical linearization. In contrast to the CADET-covariance technique, where the whole state equation is statistically linearized, a combination of both methods can be applied, which is mathematically easier to use and more efficient to compute. Moreover, this combination allows an easy design of modular computer programs structured according to physical subsystems (Figure 16) [30].

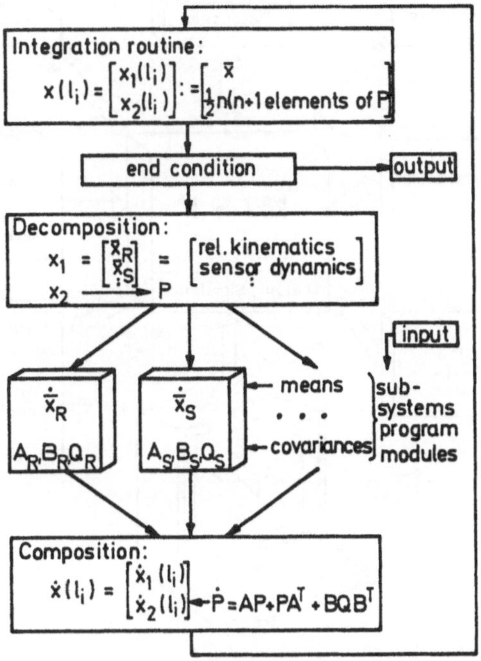

Figure 16: Program sequence of modular statistical analysis program STANHP [30] which uses Taylor linearization and statistical linearization. Output = means and covariances.

Covariance analysis via statistical linearization again is based on the assumption that the system can be suitably linearized. Furthermore, this technique is computationally advantageous in comparison to a Monte Carlo simulation, only for systems of moderate size (order < 100) [31]. Hence Monte Carlo simulation is in many cases the only way to assess nonlinear stochastic system performance. For this reason, a modular interactive Monte-Carlo missile simulation program (MISI) has been developed [34] as a flexible simulation frame work to deal with complex nonlinear stochastic guidance and control problems (Figure 17).

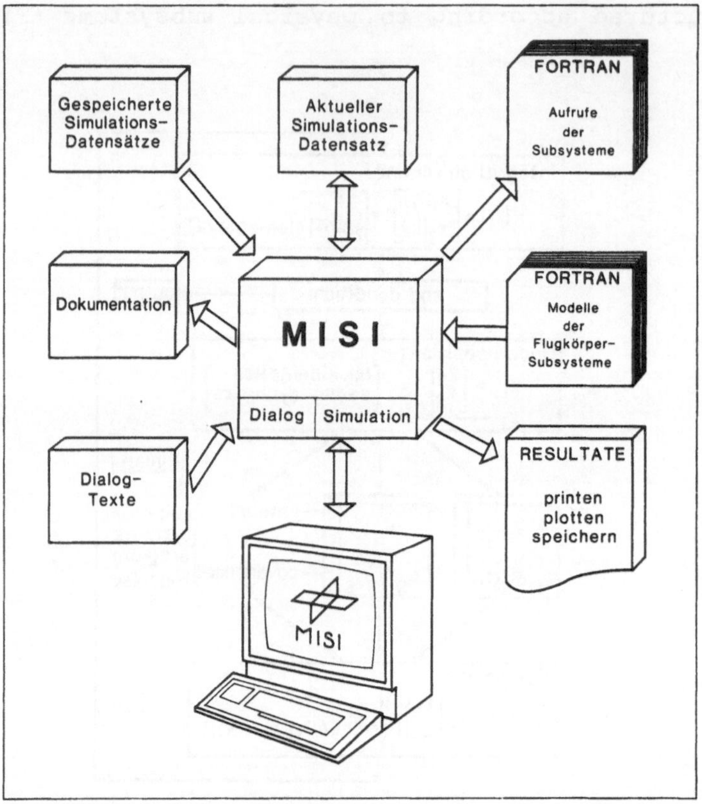

Figure 17: Information structure of the interactive Monte Carlo simulation program MISI for missile guidance analysis [34].

UNCERTAINTY IN OPERATIONAL STATE BECAUSE OF INFORMATION FAILURES

Uncertainty in the operational state may be related to two levels of information failures: sensor failures in automatic control loops and information-display failures in higher level man-machine supervisor control loops. In both cases, information vital to proper control of the system may be completely lost or wrong information may lead to unforeseen dangerous control actions. To cope with this kind of *information failure* one needs

- *reliable diagnosis schemes and redundancy,*
- *reliable sensor and information processing technology,*

- *means to recover or dispense with information.*

Sensor diagnosis via analytic redundancy

Reliability is a basic requirement of present day and future flight control systems. It is particularly true that the most attractive control concepts, such as artificial stability for the enhancement of maneuverability and flight economy, require control systems of extremely high reliability. Traditionally this requirement is achieved by using multiple devices in the vital parts of the control systems. For example, it is necessary to triplicate the hardware and to add a majority voting mechanism in order to achieve a fail-operational capability.

However, it is obvious that the conventional hardware redundancy has many disadvantages due to cost, weight, volume, energy consumption, failure rates and maintenance costs. Therefore it is reasonable to look for alternative methods which reduce the necessary efforts in hardware without loss of reliability.

For the sensor part of a flight control system, a starting point for the reduction of hardware is the observation that the signals which have to be measured are output signals of one single plant. The plant itself is given by the aircraft motion described by the flight mechanics equations. Hence the plant outputs are not independent from each other but are internally coupled. These relations given analytically can be used for reliability purposes. Thus the hardware redundancy can be replaced by the so-called analytic redundancy.

The basic tools for utilizing the analytic redundancy are filters and observers. Their algorithms have to be implemented in the signal processing part of the control system. Thus sensor hardware is replaced by computer software.

Based on these ideas, a concept for a duplex sensor configuration supported by deterministic observers has been developed (Figure 18) which achieves the fail-operational capability of a conventional triplex system. The concept was applied to a flight control system of a transport aircraft. Flight tests have shown that in principle the failure detection concept is feasible with ordinary sensors. Only certain failures of the altitude sensor are undetectable due to the fact that the altitude has to be considered as the state of a free output integrator. As far as the operational performance of the detection concept is concerned it can be reported from the flight tests that the performance was at least good enough not to upset the pilots [36].

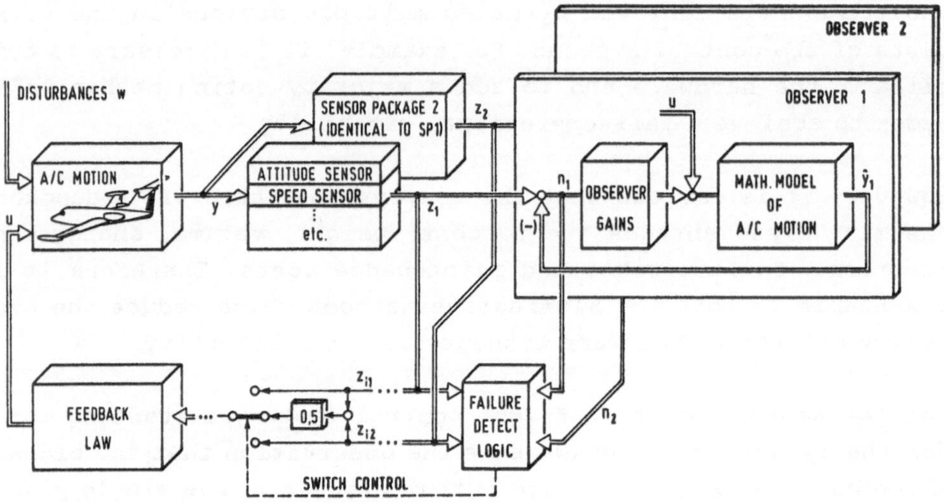

Figure 18: Flight control system with a fail-operational duplex sensor configuration based on analytic redundancy [36].

New sensor technology

Complementary to the development of reliable failure detection schemes, the DFVLR is making efforts to improve sensor reliability itself via new sensor technology. Laser gyros are an example of this kind of emerging technology. Since no wearing or rotating mechanical parts are involved, laser gyros have a long durability and are robust against vibrations and other environmental effects. At the DFVLR-Institute of Flight Guidance, a modular low-cost laser gyro (Figure 19) is currently under development [39].

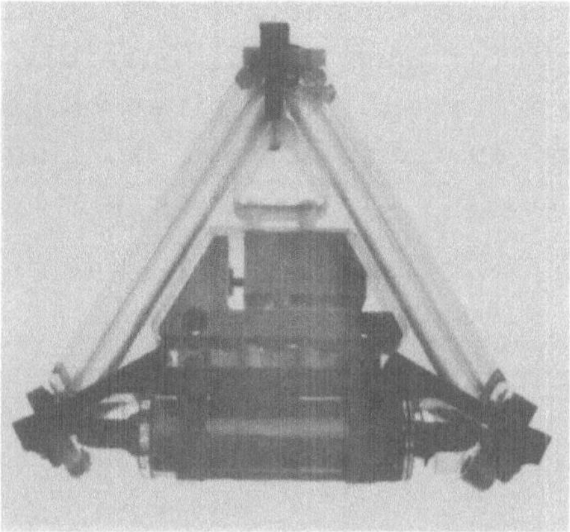

Figure 19: Modular laser gyro [39].

How to behave if an integrated information display fails?

Figure 20 shows an aircraft cockpit with a conventional and with an advanced integrated information display. The pilot receives integrated information, rather than having a great number of individual instruments yielding direct information about the state of the aircraft. This reduces his workload as supervisor and prevents decision errors as flight manager through automatic information preprocessing.

But what happens if part of the integrated display breaks down or if a "phantom error" (the most frequent occurring of all rare errors) occurs and the display is completely "dead" even for a short moment? Can the pilot be sure that after recovery the display shows the correct state, or are important pieces of previous information lost?

There are two approach philosophies to deal with this uncertainty problem [41]: In the Airbus A 310 transport aircraft the pilot is provided with an information retrieval system where he has the means to check primary information and to assure himself of the actual aircraft state of performance. In case of failure in the Boeing B 757 transport aircraft, the pilot gets adequate information about possible loss of performance and is advised to fly according to this performance level, possibly relying on lower performance back-up systems. In the first case one relies more on information back up systems, whereas in the second case one relies more on the ability of the pilot to fly an aircraft with reduced performance.

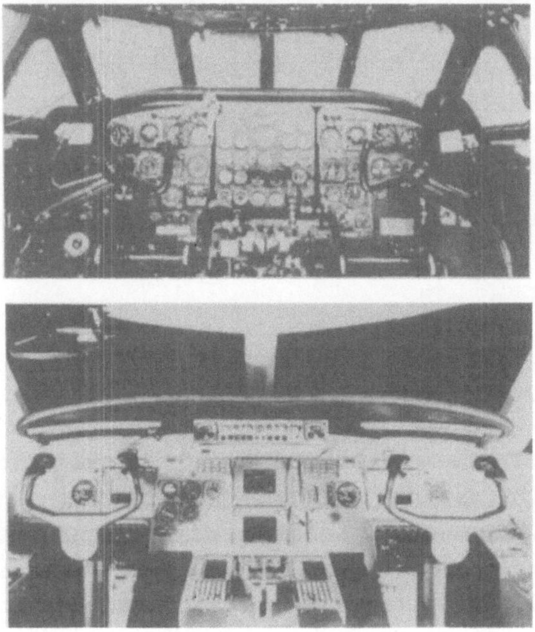

Figure 20: Conventional and advanced integrated cockpit [40].

UNCERTAINTY IN HUMAN OPERATOR BEHAVIOUR

Human operators perform functions with different levels of authority. The operator may act as a supervisor to manage slow "situation dynamics" with the possibility of emergencies, as in the case of an air traffic controller. Or, like the fighter pilot, the operator may serve as an in-the-loop controller to command and stabilize fast "motion dynamics". However, there is always some uncertainty in the way a human operator behaves in a particular situation.

The best one can do, is to take precautions against operator mistakes by selecting and training adept personnel and by conceiving and realizing the technical systems to "please" the "average" operator. In both cases one needs *fuzzy set modelling* and fuzzy logic to describe

- *"behaviour models"* for psychological selection tests and for specification of ergonomic load limits,
- *"expectation models"* like handling qualities criteria to assess man-machine performance in the design and evaluation stage,
- *"functional models"* like "paper-pilots" [11] to be used for instance in computer-aided feedback loop optimization procedures.

The DFVLR-Institute for Flight Medicine and the DFVLR-Institute for Flight Mechanics are actively involved in this field of modelling. As an example, at the Institute for Flight Medicine, highly efficient psychological pilot selection tests for Deutsche Lufthansa and air-traffic controller selection tests for the German Federal Air Traffic Control Authority have been developed. Also the European Spacelab Program has profited from this experience [42].

HOW TO DEAL WITH UNCERTAINTY?

We have seen various kinds of "uncertainty". How should it be dealt with?

First, we should try to *describe our uncertainty adequately* via

- *deterministic models* which represent typical manifestations of a disturbance effect or situation,

- *stochastic models* which characterize the randomness of a disturbance effect or situation,

- *fuzzy set models* which are able to characterize effects or situations which cannot be crisply defined in terms of set membership, but rather allows objects to have grades of membership and hence may be particularly well suited to the modelling of human behaviour which is imprecise but not random.

This modelling task is complicated by a necessary tradeoff between engineering convenience and physical correctness.

Having modelled uncertainty, how can we reduce the effects of uncertainty on system performance?.

We cannot expect good performance if we have too much uncertainty in the operational state of our system since any decision and control based on wrong information may lead to unacceptable results. Also, we cannot expect good performance if we have too much uncertainty in the operator's action in a particular situation. In this respect, we can only *try to avoid* such situations through

- *redundancy and reliability,*

- *'fail-safe' human engineering based on fuzzy set models and*

- *artificial intelligence to recover from unexpected situations.*

In case of system parameter uncertainty and external disturbances, there is a much better approach. Using the principle of *error feedback control* we may achieve *good performance in spite of*

- *system parameter uncertainty* and

- *external disturbances.*

More about feedback control will be said in the following sections.

PRINCIPLE OF ERROR FEEDBACK CONTROL

The principle of *error feedback control* consists of two elements:

- *compare the actual dynamic system response with a computed or prespecified 'nominal' response to obtain the "error" and*

- *act on the system in such a way that this error tends to be small.*

Referring to Figure 21 this means that the feedback control signal u_c should keep the error small by *counteracting* the *effects* of disturbances and parameter variations.

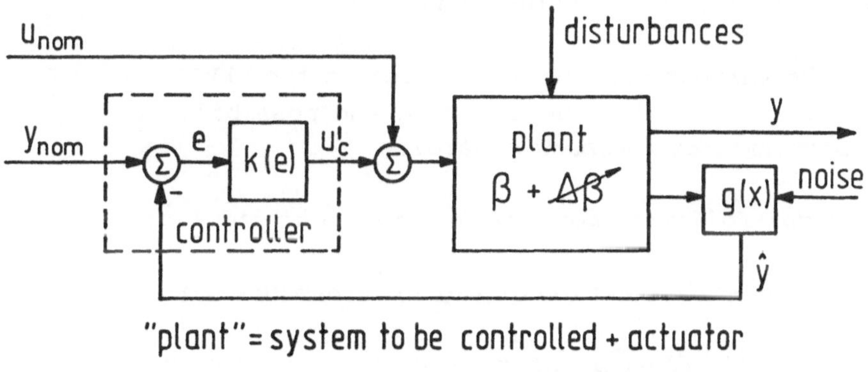

Figure 21: Principle structure for error feedback control.

The basic *control design problems* can thus be reduced to two questions:

What is to be measured?

$e = (y_{nom} - y)$: $y = g(x)$ + noise!
$g = ?$: accuracy?
cost?
reliability?

How is the control law to be chosen?

$u_c = k(e)$ $k = ?$: $\|e\|$ small!
stability margin!
no control signal saturation due to noise!

$k = f(\Delta p)$? : gain-scheduling?
adaptive?

SOME APPLICATIONS OF FEEDBACK CONTROL

From the various control activities at the DFVLR, now a few examples are selected to demonstrate the possibilities, as well as the problems, of *feedback control:*

- *model-following control for in-flight simulation*
- *robust stabilization of high-performance aircraft*
- *aircraft flutter stability augmentation via active mode decompling*
- *active damping of mechanical light-weight structures based on finite-element modelling.*

Model following control for in-flight simulation

In-flight simulators (Figure 22) are built to provide exact vision and motion cues for the investigation of complicated interactions between pilot and aircraft in demanding flight situations such as landing or when system failures occur.

Figure 22:
DFVLR In-Flight Simulators
1) HFB 320, 2) ATTAS,
3) BO 105 S3.

In-flight simulators are "variable-stability aircraft" augmented with a control system which enables the dynamics of the aircraft to be matched with a "model aircraft" to be simulated. Theoretically this could be achieved by a feedforward model-following control concept as shown in Figure 23.

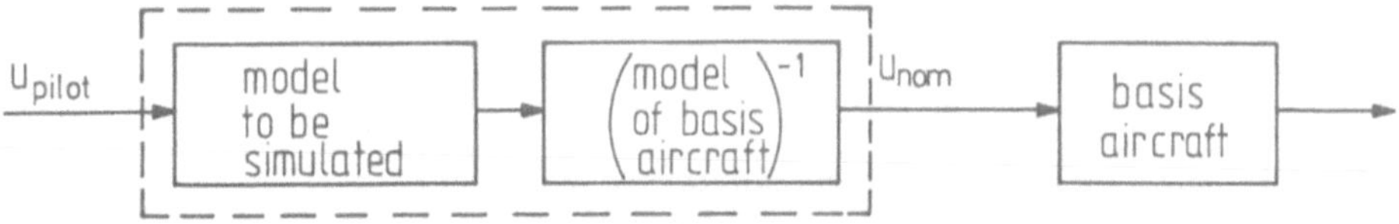

Figure 23: Feedforward model-matching control.

Such model-following control assumes an exact dynamic model of the basis aircraft, no external disturbances and sufficient control inputs such that the so-called "model-matching condition" can be satisfied. This means that a dynamic system comprising the inverse model of the basis aircraft together with the model to be simulated, is physically realizable as control law.

In practice all three conditions are not satisfied: the system parameters are not known exactly, some kind of wind disturbances must always be expected, and due to expense there is no direct side force control installed which means that only five of the six aircraft motion degrees of freedom can be matched exactly.

The first two practical restrictions can be coped with using additional feeedback control, leading to a so-called two-degrees-of-freedom control structure (Figure 24).

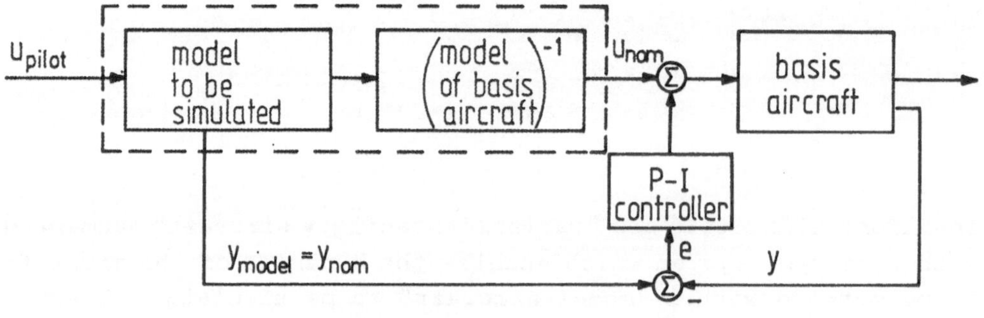

Figure 24: Two-degrees-of-freedom model following control structure for in-flight simulation; y = complete aircraft motion state.

With this control concept high accuracy model following has been achieved with the in-flight simulator HFB 320 [44]. The feedforward control part is based on a suitably simplified linear aircraft model to satisfy the model-matching conditions for five degrees of freedom of aircraft motion. The feedback part takes care of the errors due to the real (nonlinear) aircraft behaviour and external disturbances. A fixed-gain Proportional-Integral (PI) state feedback turns out to be sufficient if properly designed.

Lessons to be learned from this control example are [46]:

- *use anti-wind up for PI-control,*

- *nonlinear actuator effects have to be considered explicitly in the design, especially control saturation; the required control rate should be kept as small as possible and, a "worst case" control delay should be assumed for design,*

- *avoid to control dynamic properties which have not been modelled, e.g. high frequency actuator or structural dynamics.*

The same lessons concerning linear feedback control design are to be learned in the case of the DFVLR Helicopter In-Flight Simulator [47], where a *nonlinear* feedforward model matching technique is applied.

Robust stabilization of high-performance aircraft

Fly-by-wire technology has made it possible to define the architecture of an aircraft at the design stage without making allowances for the constraints imposed by stability requirements. The architecture of the aircraft can be optimized from the point of view of performance, economy, ease of operation, etc. by relying on the use of automatic control systems to restore aircraft stability for flight conditions in which it proves to be insufficient. However, to be able to add "artificial stability" via automatic control systems to an otherwise unstable aircraft, is not only a remedy that relieves a deficient aircraft design; in fact the freedom to shift the centre of gravity behind the aerodynamic centre, which makes the uncontrolled aircraft unstable, has several advantages.

For instance the position of the aerodynamic centre in supersonic conditions is more to the rear than in subsonic flight. Flying supersonically with a centre of gravity which provides stability in subsonic conditions would require the centre of gravity to be very much ahead of the supersonic aerodynamic centre. This in turn would lead to high deflections of the pitch control surface at supersonic speeds, and consequently, to prohibitive drag. A solution consists in centering the aircraft so as to optimize balancing in supersonic conditions (centre of gravity close to the supersonic aerodynamic centre) and in relying on the artificial stability system when flying subsonically. This solution, moreover, provides an advantage with regard to approach speeds.

What price has to be paid for the benefits? Guaranteed reliability of the stabilizing automatic control system is obviously a crucial demand in pursing the artificial stability concept since the safety of the aircraft and the success of its mission will depend essentially on the correct functioning of this system. How can this be achieved?

For the new swedish JAS 39 high-performance aircraft (Figure 25) the idea has been adopted that functional simplicity is something that should be required beforehand in order to properly guarantee reliability for the stabilization unit. Therefore the stabilization must be achieved by an automatic control system whose structure is as simple as possible. This means that

- only pitch rate q is to be used for feedback with no other aerodata or state variable measurements being employed; that is

- no control scheduling ('gain'-scheduling) as a function of the flight condition is to be used, thereby avoiding a functionally complex system which would have to be realized by software in a digital computer; instead

- an analog, that is, a hardware realization of the stabilizing control law is required with a dynamic feedback compensator and prefilter of lowest possible order and parameters independent of the flight condition.

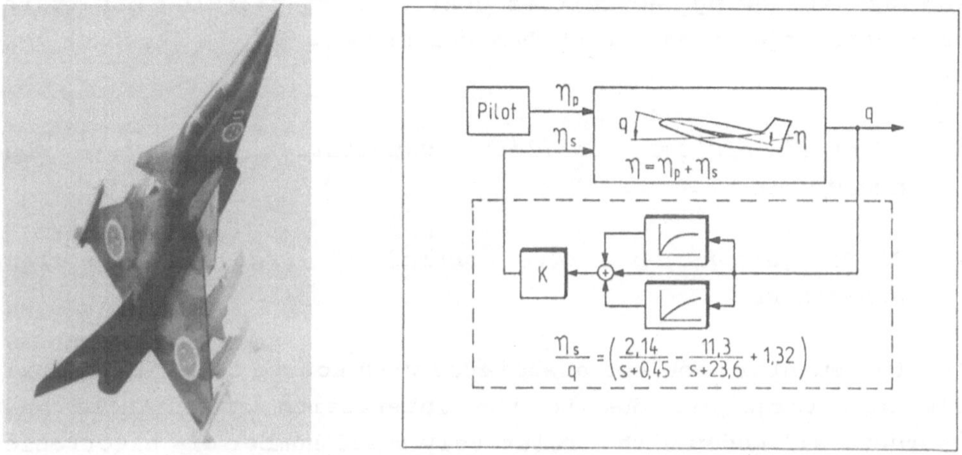

Figure 25: JAS 39 and fixed-gain feedback which stabilizes the aircraft in the entire flight envelope.

Is it possible to stabilize the aircraft under such constraints? The answer is yes [50]: A fixed gain, linear, single loop, third order dynamic controller with a "one and a half" degrees of freedom control structure can stabilize the aircraft over the entire flight envelope with remarkably good flying qualities: Pilot in-the-loop simulations with various gust levels yielded high ratings such as "the pilots like it".

Lessons to be learned from this feedback control example are:

Simple, linear, constant-gain feedback control laws can yield good performance if during design all "critical" cases concerning system parameter variations and disturbances are considered

- *individually via proper models and criteria and*

- *simultaneously over the entire range of interest.*

This obviously requires the availability of computer-aided design procedures to handle simultaneously a great number of (nonlinear) models and criteria of various kind [48,64]. Such design procedures have been developed at the DFVLR-Institute for Flight-Systems Dynamics.

The two following activities deal with stabilizing structural oscillations of *flexible mechanical structures* via feedback control:

- *Aircraft flutter stability augmentation via active mode decoupling*
- *Investigations on active control of large flexible space structures.*

The fundamental problems associated with active control of flexible structures are due to the interaction of infinite many 'structural' modes with a relatively small number of 'electronic' modes of the control system. This leads to the *questions of*

- *finite dimensional modelling and "spill over"*
- *sensor-/actuator positioning,*
- *appropriate compensator structure.*

Aircraft flutter stability augmentation via active mode decoupling

Flutter is (primarily) caused by a coupling of two distinct eigenmodes such as the 1st bending and the 1st torsional mode. For fighter aircraft, the flutter behaviour is optimized with respect to the different external store configurations defined at the design stage. However, since most of the external store configurations are specified after the aircraft is already in service, it is of no surprise if some of these "later" configurations have a restricted flight envelope due to flutter. For these particular configurations the implementation of an Active Flutter Supression System (AFSS) (Figure 26) which provides proper mode decoupling would be highly beneficial to the aircraft performance.

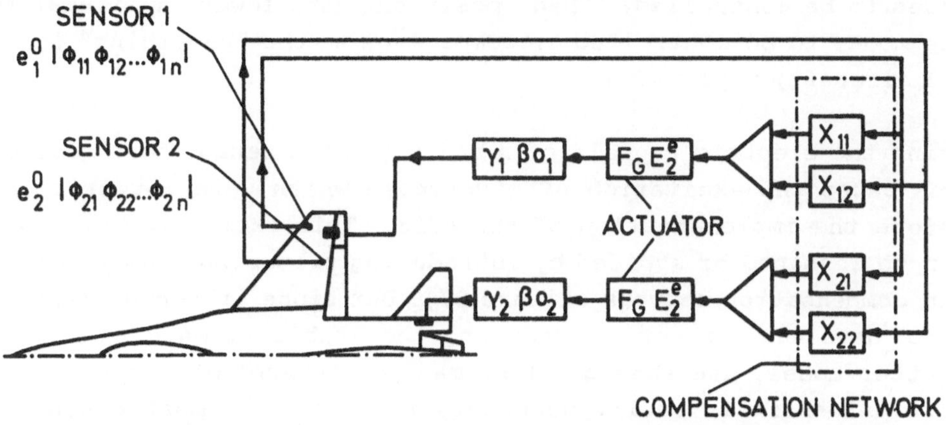

Figure 26: Active flutter suppression system layout [55].

Practical application has shown that problems arise when making use of textbook control design techniques such as those based on optimal regulator control theories. It often becomes difficult to (fully) understand the interaction effects between the passive aircraft structure and the implemented AFSS. It is especially true that the engineer involved in structural dynamics may have difficulties when he has to interprete the dynamic structural behaviour of the active controlled aircraft. Therefore, a new method for the design of active flutter suppression and gust load alleviation systems based on structural engineering insight shows great promise:

This method [55], developed at the DFVLR-Institute of Aerolasticity, allows interpretation of a basic aircraft configuration fitted with an active control system as a structurally modified configuration. Once the locations of the sensors (and actuators) are fixed, the transfer behaviour of the feedback compensation network (with regard to the amplitude and the phase) is well defined by a comparison of the controlled aircraft behaviour with a corresponding structurally modified configuration. This method is kind of an implicit model feedback technique.

A crucial part is choosing the appropriate sensor locations. The difference between "good" and "bad" sensor locations is that "good" positions are characterized by larger amplitudes in the modes to be controlled. "Bad" positions have lower amplitudes in the modes to be controlled in comparison to the "remaining" (global aircraft) eigenmodes.

Using the electric signals of a badly located sensor for feedback can cause the excitation of eigenmodes which were not critical before the implementation of the AFSS. This excitation of modes can be lessened or avoided by introducing filtering elements into the compensation network (Figure 27). But since filtering devices also have the effect of reducing the stability margins in the flutter modes, use should not be made of (sharp) electronic filters unless it is really necessary to do so. In most cases the choice of well located sensor positions greatly reduces the necessity to make use of excessive electronic filtering.

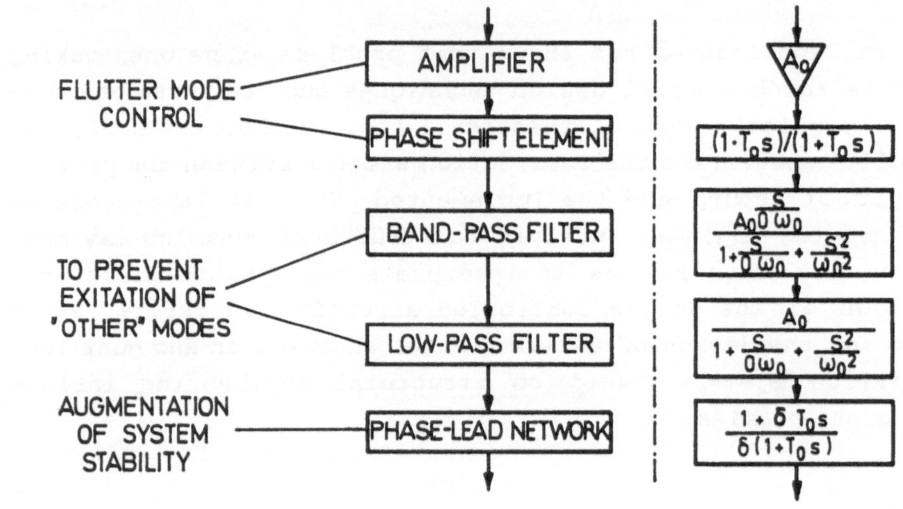

Figure 27: Elements of feedback compensator for active flutter suppression [55].

The interrelation between active control system and structural parameters helps the engineer to better understand the interaction effects of the active control system with the (flexible) aircraft structure. This clear interpretation of the interaction effects is also necessary to overcome psychological objections against the implementation of active flutter supression systems into modern aircraft. The encouraging results obtained during wind tunnel tests prove the practical applicability of this approach.

Lessons to be learned from this control example are:

- *the generic structure of a control law should be developed based on physical problem insight,*

- *additional filtering to avoid excitation of unmodelled (high frequency) modes ("spill over") may be limited by the requirement of a sufficiently large stability phase margin.*

Investigations on active control of large flexible space structures

Advanced spacecraft systems are becoming increasingly complex with large light weight structural elements being integral parts of the configuration. Furthermore, future systems are being planned which will no longer have a central rigid body. One essential characteristic of these space structures is given by their very low critical frequencies which fall within the attitude control system bandwidth. Hence, the question of vibration suppression by active modal control becomes fundamentally important.

Controller design for flexible spacecraft may proceed in two steps: first the modal damping of the flexible spacecraft is improved using simple output feedback controllers (low authority control). Second, the attitude controller is designed to satisfy the attitude requirements (high authority control) using reduced order models.

Positioning of sensors and actuators for attitude control is guided by engineering considerations. Starting from the structure of the spacecraft, one tries to measure directly by appropriate sensors the variables to be controlled and to install actuators such that they are most efficient and only small forces and torques act on the plant. In this way efficient superpositioned control loops result, where large masses are moved slowly (coarse control, low frequency) and small masses are moved fast (fine control, high frequency).

However, for modal damping of an elastic structure this methodology is not sufficient. Here, it is not the relative motion of different bodies which is to be controlled, but the relative motion of an infinite number of mass elements of a single body. The motion of the elastic structure as a continuum is described by partial differential equations which depend on internal properties like mass distribution, damping, stiffness and external boundary conditions.

The final goal in the control of elastic modes is to increase damping and/or stiffness of the structure so as to achieve the desired time behaviour. This behaviour depends on the number and locations of sensors and actuators *as well as* on the chosen structure of the controller. One therefore needs methods which allow an integrated determination of actuator/sensor positions and feedback gains for control of flexible structures. One such method [59], developed at the DFVLR-Institute for Flight Systems Dynamics together with collegues of the satellite dynamics branch, is based on the maximization of dissipation energy due to control action. The optimality criterion is determined via a Liapunov equation, and it is maximized with a recursive quadratic programming algorithm. The application of this method to a simple flexible structure (flexible beam) yields several in general dislocated actuator and sensor locations which are locally optimal. An extension of the method to treat spillover effects is implemented as an additional constraint to the optimization criterion.

Results of this design approach were experimentally verified by the DFVLR-Satellite Dynamics Branch (Figure 28) [61] using a flexible beam test set up.

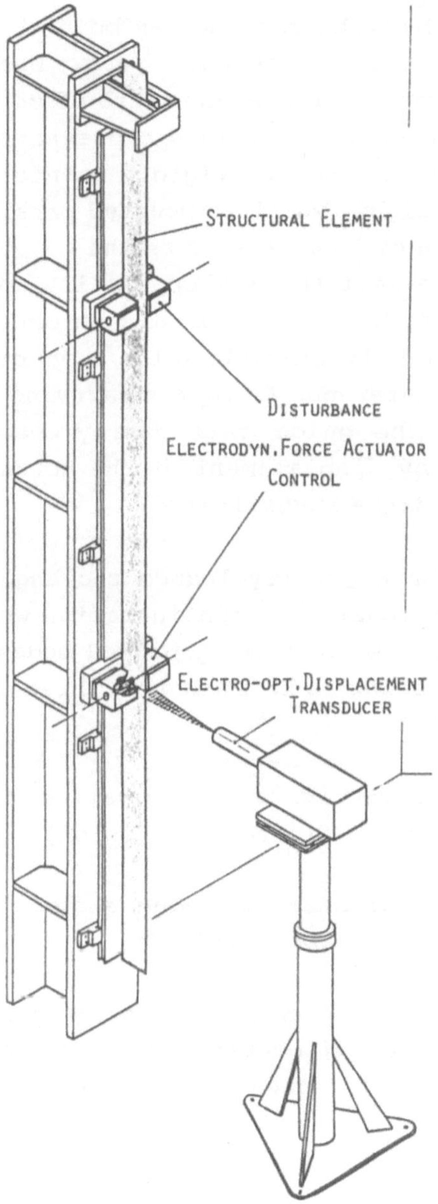

Figure 28: Flexible beam test set up [61].

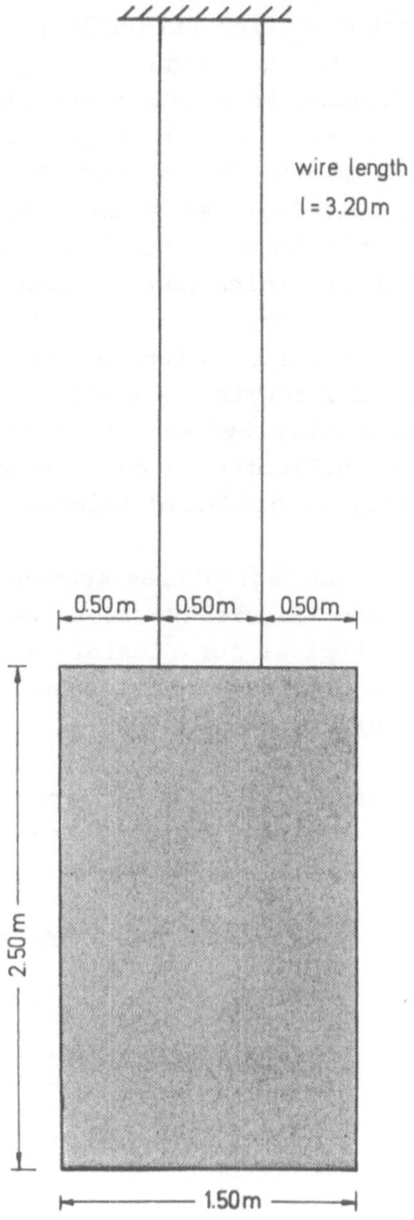

Figure 29: Vertically suspended plate.

The clamped-free homogeneous flexible beam hanging in the vertical direction (flexible pendulum) can be regarded as being typical for a large variety of flexible elements (e.g. booms, antenna) in many spacecraft systems. The beam consists of stainless steel with a length of 2.90 m with rectangular cross-section of width 10 cm and thickness 1 mm. A rigid, reinforced double-T-girder of about 500 kg weight, which is mounted parallel to the flexible beam at a distance of 5 cm, serves as the supporting mechanism and is fixed to the wall (Figure 1). At its upper end a clamp is mounted to support the flexible beam. An optical guide rail, as long as the beam, is attached along the supporting girder serving as a well-defined reference frame for carrying the actuating system. Furthermore, the guide rail easily enables repositioning to be carried out by displacement of the actuator only, if different locations need to be investigated.

Present activities concentrate on a gravity-loaded rectangular plate (Figure 29) as a two-dimensional test configuration which is typical for flexible structures with densely packed modes of low frequency. Fifteen modes are below 10 Hz including three mode clusters (Figure 30).

Lessons to be learned from these investigations (as well as from other related activities such as active flutter suppresion) are:

- *flexible structures can behave in startling ways and control laws need to be carefully verified experimentally,*

- *to avoid "spill-over", i.e. excitation of unmodelled modes, sensors and actuators should be collocated and (finite element) models of high order should be used without too drastic order reduction.*

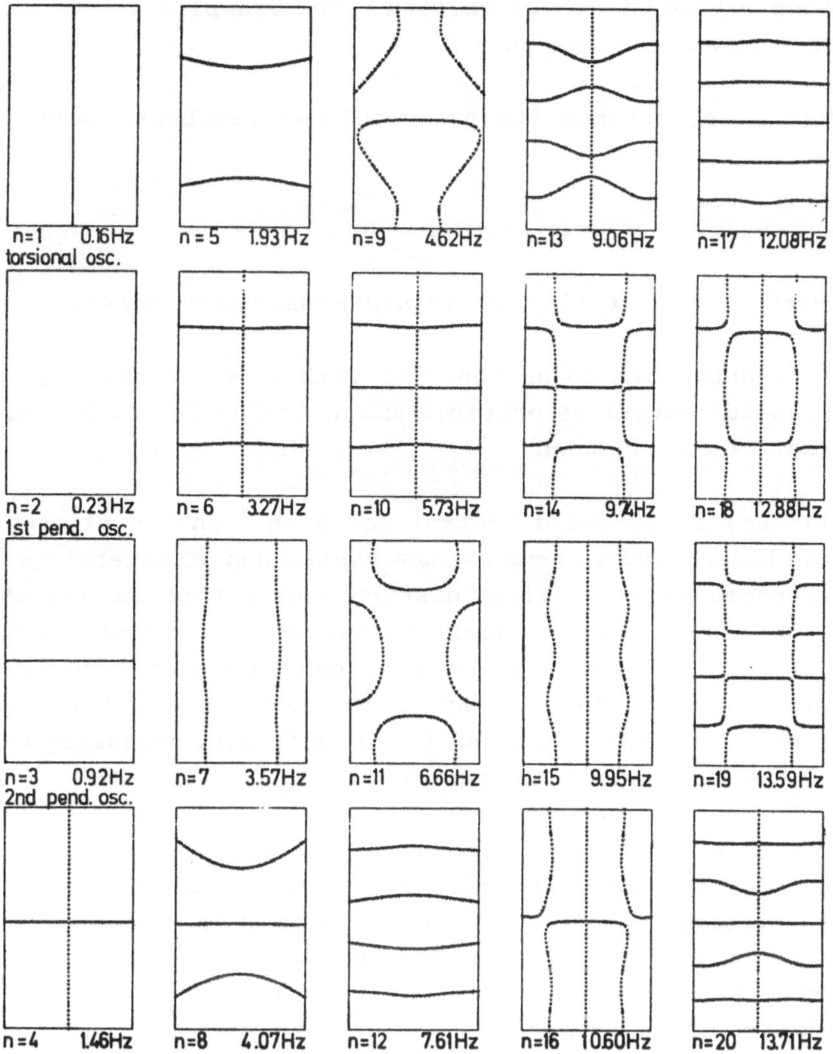

Figure 30: The lowest 20 mode shapes of the wire-suspended steel plate of figure 29 [62].

CONTROL DESIGN UNDER UNCERTAINTY: FEEDFORWARD AND FEEDBACK CONTROL

Feedback control is the only way to achieve high dynamic accuracy in spite of imprecise knowledge of the plant and in the presence of disturbance effects of the environment. The high technological achievements in the aerospace field could not have been obtained without feedback control. The examples of the preceeding section demonstrate this fact.

However, feedback has its own problems: Feedback always *introduces*

- *potential stability problems and*
- *additional disturbances through measurement noise.*

Hence feedback has to be designed with care. Actually, as much feedforward control as possible and as little feedback control as necessary should be used.

(Nonlinear) Feedforward control based on (nonlinear) plant models can be applied to command the system and to generate a "nominal" dynamic response if no disturbances act on the system. The resulting error due to modelling inaccuracies, model simplifications, and external disturbances can then be reduced by error feedback control. In this way, best use is made of the available knowledge about the plant and feedback is only necessary to counteract the effects of uncertainty.

If the error is small, it can be described by a dynamic model which is obtained by linearizing the plant model along the nominal trajectory. Vice versa, this linear error model can be used as design model to synthesise the feedback control law, -provided the resulting controller satisfies the previous assumption by keeping the error small.

The parameters of the linear design model are then a function of the operating condition and for each parameter set, i.e. each operating condition, an "optimal" controller can be designed. This suggests that the controller should be scheduled as a function of the operating condition. Such a "control scheduling" (or "gain-scheduling") is state of the art in the aerospace field. For example, aircraft controller parameters are changed as a function of the aerodynamic pressure.

However, adaptation via gain-scheduling may not even be necessary. It turns out that fixed-gain feedback, if properly designed, can often tolerate rather large plant parameter variations. To explore this possibility, the feedback parameters should not be optimized with respect to one specific plant parameter set only. Rather they should be chosen to satisfy a "best" performance compromise for a number of possible (worst-case) operating conditions. This is a so-called multi-plant-model feedback design problem. As a result, the feedback control law is good for a range of possible plant parameter sets and the extent of gain-scheduling may be reduced or it may even turn out that gain-scheduling is unnecessary.

Obviously the same idea can be applied to counteract the effects of disturbances by error feedback control. Again, the feedback control law should not be optimized with respect to one type of disturbance only. Rather, a multi-disturbance-model feedback design problem should be solved.

To solve multi-model, multi-criteria control design problems, systematic search techniques have to be used. One such technique developed at the DFVLR, is the "design systematic with vector performance index and parameter optimization" [48,64].

The general control concept of using nonlinear feedforward control based on a nonlinear plant model, and linear feedback control with gain-scheduling, is currently proposed by the DFVLR-Institute for Flight Systems Dynamics to control the DFVLR Cryogenic Wind Tunnel at Cologne (Figure 31). The nonlinear feedforward control part is generated by on-line optimization using mathematical programming, whereas the feedback controller with gain-scheduling as a function of Mach number is designed using the above mentioned design systematic with vector performance index and parameter optimization.

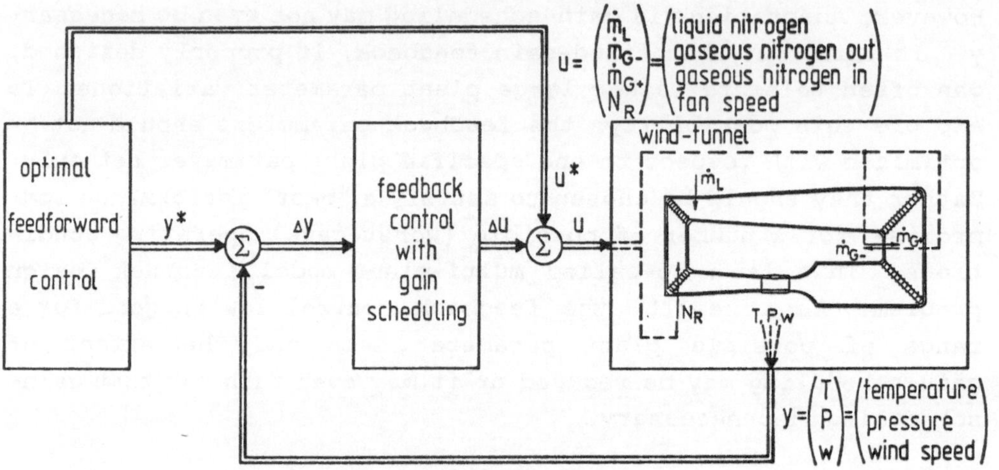

Figure 31: Multivariable two-degrees-of-freedom control structure for the DFVLR cryogenic wind tunnel.

CONCLUSION

In dealing with engineering dynamics problems one must always be aware of various kinds of uncertainty. High dynamic performance in spite of this uncertainty requires feedback control. The reported activities at the DFVLR which deal with system modelling and performance evaluation under uncertainty, as well as feedback control applications, show that successful engineering solutions are a result of merging systems and control theory with the engineering sciences of the particular field of application, e.g. flight mechanics, guidance, aeroelasticity, fluid dynamics etc. Analytic theory was always the baseline for structuring the problem solving approach. But the complexity of the technological system very often required, in addition, extensive use of computational methods such as parameter optimization and computer simulation. Last but not least, careful experimentation was necessary to verify the theoretical predictions and to assess the practicability of computer-aided engineering solutions. In all cases open-minded and engaged cooperation of experts in different fields was a prerequisite to solve the great number of grubby details which had to be settled in order to make sweet theory applicable for solving the real world problems.

ACKNOWLEDGEMENT

I thank all my colleagues at the DFVLR who developed my view on "Uncertainty and Control" by stimulating discussions and a great lot of supporting material. I hope they do not feel annoyed that many key formulations of this paper are borrowed from theirs.

REFERENCES AND RELATED DFVLR PUBLICATIONS

1 Wilhelm, K., Verbrugge, R. "Correlation Aspects in the Identification of Dynamic Effects Using Complementary Techniques", AGARD-CP-339, paper No. 17A.

2 Subke, H., Krag, B., "Dynamic Simulation in Wind Tunnels, Part II", AGARD-CP-187, 1975.

3 Hamel, P., Krag, B., "Dynamik Wind Tunnel Simulation of Active Control Systems", AGARD-CP-260, 1978.

4 Wilhelm, K., Gmelin, B., "DFVLR-Dynamic Model Testing in Wind Tunnels for Active Controls Research", 12. ICAS Congress, Munich 1980.

5 Subke, H., "Test Installation to Investigate the Dynamic Behaviour of Aircraft with Scaled Models in Wind Tunnels", Symposium on Dynamic Analysis of Vehicle Ride and Manoeuvering Characteristics, London, 1978.

6 Krag, B., "The Wind Tunnel Behaviour of a Scaled Model with a Gust Alleviation System in a Deterministic Gust Field", Symposium on Dynamic Analysis of Vehicle Ride and Manoeuvering Characteristic, London, 1978.

7 Marchand, M., "Bestimmung der Derivative eines DO-28-TNT-Modells aus Freiflugversuchen", DFVLR-FB 82-17, 1982.

8 Hamel, P., "Aircraft Parameter Identification Methods and their Applications - Survey and Future Aspects", AGARD-LS-104, 1979.

9 Marchand, M., Koehler, R., "Determination of Aircraft Derivatives by Automatic Parameter Adjustment and Frequency Response Methods", AGARD-CP-172, 1975, paper No. 17.

10 Plaetschke, E., Schulz, G., "Practical Input Signal Design", AGARD-LS-104, 1979.

11 Koehler, R., "Entwurf und Realisierung von Eingangssignalen zur Indentifizierung eines Pilotenmodells", DFVLR-FB 84-08, 1984.

12 Plaetschke, E., Mulder, J.A., Breeman, J.H. "Results of Beaver Aircraft Parameter Identification", DFVLR-FB 83-10, 1983.

13 Jategaonkar, R.V., Plaetschke, E., "Maximum Likelihood Parameter Estimation from Flight Test Data for General Non-Linear Systems", DFVLR-FB 83-14, 1983.

14 Jategaonkar, R.V., Plaetschke, E., "Non-Linear Parameter Estimation from Flight Test Data Using Minimum Search Methods", DFVLR-FB 83-15, 1983.

15 Plaetschke, E., Jategaonkar, R.V., "Maximum-Likelihood-Parameterschätzung in nichtlinearen flugmechanischen Systemen", Zeitschrift für Flugwissenschaften und Weltraumforschung 8, S. 149-154, 1984.

16 Plaetschke, E., Mulder, J.A., Breeman, J.H., "Flight Test Results of Five Input Signals for Aircraft Parameter Identification", Proceedings of 6th IFAC Symposium on Identification and System Parameter Estimation, June 1982, pp. 1021-1026.

17 Mackie, D.B., "A Comparison of Parameter Estimation Results from Flight Test Data Using Linear and Nonlinear Maximum Likelihood Methods", DFVLR-FB 84-06, 1984.

18 Kaletka, J., "Einsatz der CRAY-1S im Bereich der Simulation und Kennwertermittlung bei Hubschraubern", DFVLR WT-Colloquium, Köln, 7.11.1984.

19 Doherr, K.F., Koehler, R., "Bedarf und Nutzen der Systemidentifizierung in der Flugmechanik", DFVLR Nachrichten, H. 43, S. 12-16, Nov. 1984.

20 Adam V., Lechner, W. "A Concept for the 4D-Guidance of Transport Aircraft in the TMA", Proceedings 13th ICAS-Congress, 1982, Seattle, USA.

21 Adam, V., Lechner, W., "Investigations on Four-Dimensional Guidance in the TMA", AGARD Conference Proceedings, No. 340, Lisbon, September 1982.

22 Köpp, F. "Erstellung und Erprobung des Laser-Doppler-Anemometers zur Fernmessung des Windes", DFVLR-FB 83-14, 1983.

23 Hotop, H.J. "Anwendung der Regressionsanalyse zur Ermittlung der Luftdaten-Sensorfehler mittels eines Trägheitsnavigationssystems", DFVLR-Mitt. 84-03, 1984.

24 Hotop, H.J., Lechner, W., Stieler, B., "Probleme bei der bordseitigen Bestimmung des Windes mit Optimalfiltern", Symposium 'Fliegen im Flughafen-Nahbereich', DGON, Bestell-Nr. 120-1, Hamburg, April 1979.

25 Lechner, W., Onken, R., "Wind Modelling for Increased Aircraft Operational Efficiency", AGARD, 39th Symposium of the Guidance and Control Panel, Ankara, Turkey, 1984.

26 Lechner, W., "Algorithmen zur automatischen 4-dimensionalen Flugbahnführung unter Berücksichtigung der momentanen Windsituation", DFVLR-FB 84-40, 1984.

27 Arbter, K., "Ein analytisches Modell für simultan empfangende Radarsensoren zur Winkelmessung", DFVLR-IB-552-80/11.

28 Arbter, K., "Ein nichtlineares, stochastisches Modell eines halbaktiven Monopulssystems", DFVLR-IB-552-80/12.

29 Froriep, R., Joos, D., "STANHP - Programm zur statistischen Anylyse von Lenkschleifen für Flugkörper mit Zielsuchlenkung", DFVLR-Mitt. 80-17, 1980.

30 Froriep, R., Joos, D., "Stochastic Simulation Using Covariance Techniques: Modular Program Package for Nonlinear Missile Guidance", AIAA J. Guidance, vol. 7, p. 509-512, 1984.

31 Froriep, R., Joos, D., "Stochastic Simulation Using Covariance Techniques. A Modular Program Package for Nonlinear Missile Guidance Loops", 10th IMACS World Congr. on Systems Simulation and Scient. Computation, Concordia Univ., Montreal/Can. 8.-13.08.1982.

32 Joos, D., "Extension of Proportional Navigation by the Use of Optimal Filterung and Control Methods" ESA TT-683 (english translation of DFVLR-FB 80-10, 1980).

33 Hofmann, W., Joos, D., "Missile Guidance Techniques", AGARD LS-101 Guidance and Control for Tactical Guided Weapons with Emphasis on Simulation and Testing, 1979.

34 Somieski, G., Uhrmeister, B., "MISI - Ein allgemeines modulares und interaktives Flugkörper-Simulationsprogramm", DFVLR Nachrichten, H. 43, S. 19-23, 1984.

35 Stuckenberg, N., "An Observer Approach to the Identification and Isolation of Sensor Failures in Flight Control Systems", ESA TT-738 (english translation of DFVLR FB 81-26, 1981).

36 Stuckenberg, N., "A Diagnosis Scheme for Sensors of a Flight Control System Using Analytic Redundancy", AGARD CP-349 Integration of Fire Control, Flight Control and Propulsion Control Systems.

37 Stuckenberg, N., "Optimal Dectection of Sensor Failures in Flight Control Systems Using Deterministic Observes", in AGARDOGRAPH 289 Fault Tolerant Considerations and Methods for Guidance and Control Systems.

38 Stuckenberg, N., "The application of observes to the identification of Sensor Failures in Control Systems", Proceedings 1st Symposium on Applied Control and Identification - ACI'83, Kopenhagen, 1983.

39 Rodloff, R., "Laserkreisel mit optimierter Resonatorgeometrie", DFVLR-Nachrichten, H. 43, S. 24-28, 1984.

40 Thomas, F., "Flugführung und Flugsicherung - Gegenwärtige und zukünftige Aufgaben für Forschung und Industrie", DFVLR Nachrichten, H. 24, S.2-9, 1978.

41 Beyer, R., persönliche Mitteilung.

42 Steininger, K., "Neues Auswahlsystem für Flugsicherungspersonal", DFVLR-Nachrichten, H. 43, S. 35-37, 1984.

43 Wilhelm, K., Schafranek, D., "In-Flight Investigation of the Influence of Pitch Damping and Pitch Control Effectiveness on Landing Approach Flying Qualities of Statically Unstable Transport Aircraft", DFVLR-FB 84-12, 1984.

44 Lange, H.-H., "Flugerprobung des Modellfolgereglers für die HFB 320 zur Simulation des Airbus A 310 im Fluge", DFVLR-Mitt. 79-13, 1979.

45 Hanke, D., "In-Flight-Simulation in der Flugmechanik", DFVLR Nachrichten, H. 43, S. 5-11, Nov. 1984.

46 Henschel, F., "Über die Berechnung von Regelungen zur In-Flight Simulation unter Berücksichtigung von Stellglieddynamik", DFVLR-FB 85- , 1985.

47 Hilbert, K.B. Bouwer, G., "The Design of a Model-Following Control System for Helicopters", Proc. AIAA Guidance and Control Conference, p. 601-617, Seattle, 1984.

48 Kreisselmeier, G., Steinhauser, R., "Application of Vector Performance Optimization to a Robust Control Loop Design for a Fighter Aircraft", DFVLR-FB 80-14, 1980, reprinted Int. J. Control, 1983, 37, No. 2, p. 251-284.

49 Franklin, S.N., Ackermann, J., "Robust Flight Control: a Design Example", AIAA Journal of Guidance and Control 1981, vol. 4, 597-605.

50 Grübel, G., Joos, D., Kaesbauer, D., "Robust Back-Up Stabilization for Artificial-Stability Aircraft", Proc. 14th ICAS Congress, vol. II, p. 1085-1095, 1984.

51 Freymann, R., "Structural Modifications on a Swept Wing Model with Two External Stores by Means of Modal Perturbation and Modal Correction Methods", ESA-TT-463, 1978.

52 Freymann, R., "Nonlinear Aeroelastic Analyses Taking into Account Active Control Systems ", AGARD-R-698, 1981.

53 Freymann, R. "Aktive Flatterunterdrückung am Beispiel des Segelflugzeugs SB-11", DFVLR-IB 232 - 82J08, 1982.

54 Freymann, R., "Eine Methode zur Auslegung des Reglers von aktiven Flatterunterdrückungssystemen", ZFW Z. Flugwiss. Weltraumforsch. 7, S. 407-416, 1983.

55 Freymann, R., "New Simplified Ways to Understand the Integration Between Aircraft Structure and Active Control Systems", Proc. AIAA Guidance and Control Conference, p. 233-245, Seattle, 1984.

56 Schulz, G., Heimbold, G., "Robust Active Vibration Damping of Flexible Spacecraft", Proc. AIAA Guidance and Control Conference, p. 787-793, Gatlinburg, USA, 15.-17.08.1983.

57 Schulz, G., "Low Authority Control of Flexible Spacecraft via Numerical Optimization", AGARD Symposium on Guidance and Control Techniques for Advanced Space Vehicles, Florenz, Italien, 27.-30.09.1983.

58 Schulz, G., Heimbold, G., "Robuste aktive Schwingungsdämpfung von Finite-Elemente-Strukturen in der Raumfahrt", ZFW Zeitschrift für Flugwissenschaften 7, 1983, 2. S. 91-99.

59 Schulz, G., Heimbold, G., "Dislocated Actuator/Sensor Positioning and Feedback Design for Flexible Structures", AIAA J. Guidance, Control and Dynamics 6, 1983, 5, S. 361-367.

60 Schulz, G., Heimbold, G., "Zur Positionierung von Stellgliedern und Sensoren mit gleichzeitiger Reglerauslegung für die Regelung großer flexibler Raumfahrtstrukturen", Z. rt Regelungstechnik 31, 1983, 6, S. 188-196.

61 Schäfer, B., Holzach, H., "Experimental Research on Flexible Beam Modal Control", AIAA J. Guidance, Control and Dynamics, Juli/August 1985.

62 Schäfer, B., "Dynamical Modelling of a Gravity-Loaded Rectangular Plate as a Test Configuration for Attitude Control of Large Space Structures", 35h IAF Congress, Lausanne, 1984.

63 Schäfer, B., "Identification and Model Adjustment of a Hanging Plate Designed for Structural Control Experiments", paper submitted to 2nd Int'l. Symposium on Structural Control, Ontario, Canada, 1985.

64 Grübel, G., Kreisselmeier, G., "Systematic Computer Aided Control Design", AGARD, LS-128, Computer Aided Design and Analysis of Digital Guidance and Control Systems", 1983, S. 81-87.

SYSTEM IDENTIFICATION

Lennart Ljung
Department of Electrical Engineering
Linköping University
S-581 83 Linköping, Sweden

1. INTRODUCTION

Uncertainty is the major stumbling block in most problems of design and decision. For control design, uncertainty relates to our lack of precise knowledge about the properties of the process to be controlled as well as about the disturbances that will affect it. System identification and modelling represent a head-on approach to deal with this uncertainty problem: to construct mathematical models of the process and its environment using available information, as well as to, when possible, perform experiments that will provide further useful information.

In this contribution we shall give a brief overview of identification and modelling. We shall then concentrate on typical aspects of the area and outline basic procedures and their properties, rather than deal with specific algorithms. A discussion of the model concept is first given in Section 2 and then the System Identification problem is outlined in Section 3. Section 4 deals with time-domain analytic models in somewhat more detail, while Section 5 describes how such models can be fitted to observed data. The properties of the resulting models are summarized in Section 6 and some user aspects are summarized in Section 7. The important problem of model validation is described in Section 8. Conclusions and an outlook of the area are given in the final Section 9.

2 SYSTEMS AND MODELS

The notion of systems plays an important role in modern science. Many problems in various fields are solved in a systems oriented framework. Subjects like control theory, communication theory and operations research tell us how to determine suitable regulators, filters, decision rules etc. Such theory assumes that a model is available of the system in question. The applicability of the theory is thus critically dependent on the availability of good models.

How does one construct good models of a given system? This question about the interface between the real world and the world of mathematics is crucial for successful applications. The general answer is that we have to study the system experimentally and make some inference from the observations. In practice there are two main routes. One is to split up the system, figuratively speaking, into subsystems, whose properties are well understood from previous experience. This basically means that we rely upon "laws of Nature" and other well established relationships, that have their roots in earlier empirical work. These subsystems are then joined together mathematically, and a model of the whole system is obtained. This route is known as modelling, and does not necessarily involve any experimentation on the actual system. When a model is required of a yet unconstructed system (such as a projected aircraft) this is the only possible approach. The procedure of modelling is quite application dependent and often has its roots in tradition and specific techniques in the application area in question. Basic techniques typically involve structuring of the process into block diagram with blocks consisting of simple elements. The reconstruction of the system from these simple blocks is now increasingly being done by the computer. A survey of modelling techniques is given in Fasol and Jörgl (1980). See also Wellstead (1979).

The other route is based on experimentation. Input and output signals from the system are recorded and are then subjected to data analysis in order to infer a model of the system. This route is known as identification. It is often advantageous to try to combine the approaches of modelling and identification in order to maximise the information obtained from identification experiments and to make the data analysis as sensible as possible.

Many different types of models of dynamical systems have been developed for various purposes. We shall here give a brief list of some common ones.

Intuitive or mental models. In many cases a model of a system is never formalized. The user works with an intuitive or mental picture of how the system operates, and uses that to solve design problems associated with the systems. Such a mental model can be verbalized in a number of different ways, like, e.g. time scale for dominating time constants or frequency ranges for certain resonances etc.

<u>Graphic models</u>. A linear system is fully characterized by its impulse response or by its step response. Plots or tables of this function will thus constitute a model of the system. Such a model can be used for, e.g., tuning a PID regulator. An equivalent and more common choice is to describe the frequency response function $G(i\omega)$ as a function of ω. This function is the Fourier transform of the impulse response and equal to the transfer function $G(s)$ evaluated on the imaginary axis. The argument of the complex number $G(i\omega)$ describes the phase shift between an input sinusiod of frequency ω and the corresponding output sinusoid, while its absolute value is the ratio of output-to-input sinusoid amplitude.

A plot of the frequency response function, such as the Nyquist or Bode diagram, is a very common model of a linear dynamical system. Several control design techniques have been specifically tailored to such models.

<u>Analytic models</u>. For many purposes it is more advantageous to work with analytic, mathematical models, where the relationships between input signals $u(t)$ and output signals $y(t)$ are described by mathematical expressions. Most often, these are basically differential equations, ordinary or partial, simple because most physical phenomena are usually described in that way. A linear model is then

$$y^{(n)}(t)+a_1y^{(n-1)}(t)+a_2y^{(n-2)}(t)+\ldots+a_ny(t)=b_1u^{(n-1)}(t)+\ldots$$
$$\ldots+b_nu(t) \qquad (1)$$

where (k) as superscript denotes k times differentiation.

It is also common to work with models in discrete time. The counterpart of (1) then is

$$y(t_k)+a_1y(t_{k-1})+\ldots+a_ny(t_{k-n})=b_1u(t_{k-1})+\ldots+b_nu(t_{k-n}) \qquad (2)$$

Here $\{t_k\}$ are the sampling instants. Formulas for how to transform from (1) to (2) when the input is constant between the sampling instants are given, e.g., in Kwakernaak and Sivan (1972).

Often the effects of random disturbances on the system and on the measurements are included in the model. A typical example is a stochastic, state space model

$$x(t_{k+1}) = F\,x(t_k) + G\,u(t_k) + w(t_k)$$
$$y(t_k) = H\,x(t_k) + e(t_k) \tag{3}$$

where w and e are white noise sequences with certain specified covariance properties.

Analytical models are most often given in the time domain, but this is no inherent feature. We could have a frequency domain, analytical model like

$$G(i\omega) = \frac{b_1(i\omega)^{n-1}+b_2(i\omega)^{n-2}+\ldots+b_n}{(i\omega)^n+a_1(i\omega)^{n-1}+\ldots+a_n} \tag{4}$$

which is equivalent to (1).

The list of possible analytical models can be made long and it would lead too far here to try and make it complete. In Section 4 we shall return to the subject.

3 AN OUTLINE OF SYSTEM IDENTIFICATION

Approaches to identification

The identification procedure can in general terms be described as follows.

1. Collect input-output data from the process.

2. Settle for a set of candidate models.

3. Pick one particular member of the model set as the best representative, guided by the information in the data.

Let us give a conceptual discussion of each of these three steps. More details will follow below.

1. The data. The data is sometimes recorded during a specifically designed identification experiment. The objective then is to get as good

information about the system as possible. Some methods to determine models require special input sequences, and thus specific experiments. In other cases data from normal operation of the system have to be used.

Nowadays, data is almost always recorded by sampling in discrete-time using a digital computer (perhaps after an intermediate storage on a tape with a data recorder). We shall here throughout assume that this is the case and we shall denote the data set recorded over N samples by Z^N.

2. The set of models. A set of candidate models is obtained by specifying among which collection of models we are going to look for a suitable one. For graphic models this will typically be the set of all (reasonably smooth) curves - corresponding to the set of all linear models in the cases described in the previous section. For analytic models suitable sets are usually obtained by letting certain parameters in the model descriptions range over a given space (such as the a_i and b_i of (2) or certain entries of the matrices F, G, H in (3)). We shall generally denote a set of models by $\mathcal{M}$.

A model set whose parameters are basically viewed as vehicles for adjusting the fit to data, and do not reflect physical considerations in the system is called a black box model (set). Model sets with adjustable parameters with physical interpretations may, accordingly, be called grey boxes.

3. Picking a particular model, in $\mathcal{M}$, guided by data. This is "the identification method". There is obviously a vast number of ways to select models. To give an idea about basic principles, it is useful to distinguish between methods for graphic models and for sets of analytic models.

Graphic time domain-models: Graphic time-domain models are basically impulse or step responses. Techniques to determine these are called transient analysis. One simply applies an (approximate) impulse or a step as input and records the corresponding output. If the signal-to-noise ratio is good for the measurements, valuable information about static gain and dominating time constants can be obtained in this manner. There are certain techniques to approximate given step responses with analytic, low order models, based on the tangent with the largest slope; see Rake (1980). The impulse response can also be obtained as the crosscorrelation function between the output and the input, when the input sequence is white noise, see, e.g. Godfrey (1980). Apparatus that performs such correlation analysis is commercially available.

Graphic frequency domain models: A direct way of determining the value of the frequency response function of a system at a given frequency, is to use a sinusodial input of that frequency, let the transient die out, and record the phase shift and amplitude change of the output sinusoid. Formally, in discrete time (T=sampling interval) we have

$$u(t) = u_0 \sin(\omega kT) \qquad kT \leq t < (k+1)T \tag{5}$$

$$y(kT) = y_0 \sin(\omega kT+\varphi) + \text{transient} \tag{6}$$

The frequency response function $\hat{G}(e^{i\omega})$ is then determined from

$$\arg \hat{G}(e^{i\omega}) = \varphi; \qquad |\hat{G}(e^{i\omega})| = y_0/u_0 \tag{7}$$

The experiment is repeated for a number of different frequencies ω in the interesting range, and a table or graph of $G(e^{i\omega})$ can be constructed. This technique is known as frequency response analysis. See Rake (1980).

If the measurements are noise corrupted so that it is difficult to determine y_0 and φ directly, it is useful to correlate out the periodic component by multiplying $y(kT)$ by $\sin(kT\omega)$ and by $\cos(kT\omega)$, respectively, and sum over a number of observations. The phase shift and amplitude gain can then be determined more accurately. The technique is known as frequency analysis by the correlation methods, and equipment for this is commercially available.

With more sophisticated data analysis one may, so to speak, apply all frequencies at the same time and sort them out afterwards by Fourier techniques. This gives a frequency response function

$$\hat{\hat{G}}_N(e^{i\omega}) = Y_N(\omega)/U_N(\omega) \tag{8}$$

$$Y_N(\omega) = \frac{1}{\sqrt{N}} \sum_{t=1}^{N} y(t)e^{-i\omega t}$$

$$U_N(\omega) = \frac{1}{\sqrt{N}} \sum_{t=1}^{N} u(t)e^{-i\omega t}$$

where Y_N and U_N are the discrete Fourier transforms (DFT) of the output and input sequences using N data. When noise affects the systems (8) is usually a bad estimate ("the periodogram estimate"), since no noise reduction is obtained. Instead various smoothing filters are applied to (8): weighted averages over certain frequency windows are formed. Such techniques are known as spectral analysis and are further described in e.g. Jenkins and Watts (1968) and Brillinger (1981) The estimate can thus be written

$$\hat{G}_N(e^{i\omega}) = \int_{-\pi}^{\pi} W_\gamma(\zeta-\omega)\hat{\hat{G}}_N(e^{i\zeta})d\zeta \tag{9}$$

where $W_\gamma(\zeta)$ is a weighting function that typically has its "mass" concentrated arouind $\zeta=0$. See also Ljung (1984).

Analytical models. Most modern identification methods deal with the estimation of analytical models, usually described in the time domain. More detailed descriptions of such methods will be given in Section 5. See also Goodwin and Payne (1977), Åström (1980) and Åström and Eykhoff (1971). The basic idea behind such methods is the following one: Let each of the candidate models "guess" (predict) the next output y(t) based on the information in Z^{t-1}. Pick that model that produces the best ("smallest") sequence of errors between guesses and actually recorded outputs. These identification methods are thus characterized by a criterion of fit between a model and the recorded sequence of data.

The Identification Procedure

The identification procedure is, from a principal point of view, described by the three items listed above: Data, Model set, and Identification criterion. From a practical point of view the procedure is characterized by a number of choices which we now list.

Experiment Design. How to design the identification experiment, so that it becomes suitably informative. Choice of inputs, sampling rates, presampling filters, feedback configurations, choice of signals to be measured, etc. Such aspects are further discussed in Goodwin and Pavne (1977).

Choice of Model Set. To select the set of candidate models. This is no doubt the most important and, at the same time, the most difficult choice. It is here that a priori knowledge and engineering intuition and insight has to be combined with formal properties of models and identification methods to facilitate a good result from the identification exercise.

Choice of criterion of fit. How to evaluate the quality of a particular model from data is a crucial issue, which is further discussed in Section 5.

Calculation of the best model. With given data and a fixed model set and a chosen criterion of fit, the "best" model is implicitly defined. It remains "only" to calculate it, which may involve extensive computations. Good numerical algorithms are required in order to allow for reliable and inexpensive calculations. Some computational aspects are included in Ljung (1982). See also Gupta and Mehra (1974), and Dennis and Schnabel (1983).

In several applications, the models are required on-line, as the system operates and more data becomes available. The reason could be that the models are to be used for some on-line decision like control (adaptive control), filter tuning (e.g. adaptive noise cancellation) or monitoring (fault detection). This implies certain restrictions for how to calculate the estimates. Such methods are called recursive identification methods (or on-line or real-time identification) and are discussed, e.g., in Ljung and Söderström (1983).

Model validation. Once the best model available in the model set has been determined it remains to test whether it is "good enough", i.e. whether it is valid for its purpose. This is the problem of model validation, which is further described in Section 8.

The whole identification procedure is typically an iterative one, in which earlier made choices have to be revised after the model validation step and portions of the procedure repeated. This is illustrated in figure 1.

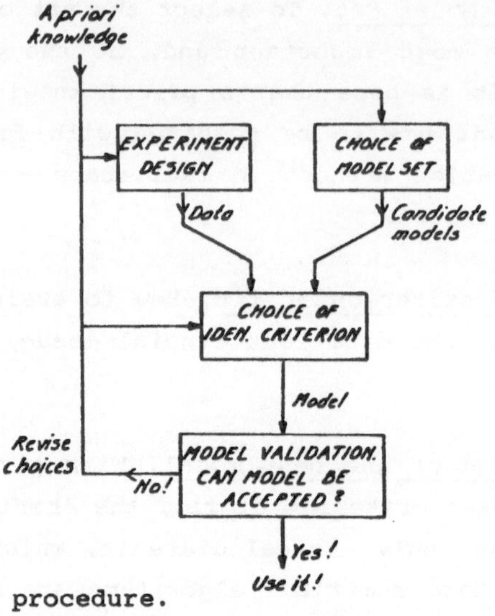

Figure 1. The identification procedure.

The identification tool

System identification has become an important tool to solve a number of modelling problems in engineering. Some aspects of the applicability of this tool to real life problems are discussed in Gevers and Bastin (1982). Here we shall briefly comment upon what this tool should look like in the hand of the user. The identification procedure is, as pointed out in figure 1, typically an iterative one, where insights and judgements of the user are mingled with formal calculations, extensive data handling and complex algorithms. To make the tool an efficient one it is therefore more or less necessary to package the software in an interactive environment, with man-machine communication via graphical displays. Several such packages have been developed. A well-known one, IDPAC, developed at the Lund Institute of Technology is described in Åström (1980) and Wieslander (1979).

4 ANALYTIC TIME-DOMAIN MODELS

Describing dynamical systems in the time domain allows a considerable amount of freedom. Usually, differential (partial or ordinary) equations are used to describe the relationships between inputs and outputs. In discrete-time (sampled-data systems) difference equations are used instead. The question of how to describe properties of various disturbance signals also allows for several different possibilities. Here, we shall list a few typical choices, confining ourselves to the case of linear, discrete time models. We shall for easy notation use the sampling interval as the time unit, thus assuming it to be constant.

The word "model" is sometimes used ambigously. It may mean a particular description (with numerical values) of a given system. It may also refer to a description with several coefficients or parameters that are not fixed. In the latter case, it is more appropriate to talk about a model set: a set of models that is obtained as the parameters range over a certain domain.

Linear difference equations. Let the relationship between the input sequence $\{u(t)\}$ and the output sequence $\{y(t)\}$ be described by (2) or with $t_k=k$,

$$y(t)+a_1y(t-1)+....+a_ny(t-n)=b_1u(t-1)+....+b_mu(t-m) \tag{10}$$

Here the coefficients a_i and b_i are adjustable parameters. (A multivariable description would be quite analogous, with a_i and b_i as matrices.) We shall generally denote the adjustable parameters by a vector θ:

$$\theta = (a_1...a_n\ b_1...b_m)^T. \tag{11}$$

If we introduce the vector of lagged inputs and outputs

$$\varphi(t)=(-y(t-1)...-y(t-n)u(t-1)...u(t-m))^T, \tag{12}$$

the equation (1) can be rewritten in the more compact form

$$y(t) = \theta^T\varphi(t). \tag{13}$$

In (10) or (13), the relationship between inputs and outputs is assumed to be exact. This may not be realistic in a number of cases. Then we may add a term $v(t)$ to (1) or (13):

$$y(t) = \theta^T\varphi(t) + v(t) \tag{14}$$

that accounts for various noise sources and disturbances that affect the system, as well as for model inaccuracies. This term can be further modelled, typically by describing it as a stochastic process with certain properties. The simplest model of that kind is to assume

$\{v(t)\}$ to be white noise, i.e. a sequence of independent random variables with zero mean values. However, many other possibilities exist. Among the most common models is the following one.

<u>ARMAX models</u>. If the term $\{v(t)\}$ in (14) is described as a moving average (MA) of white noise $\{e(t)\}$ we have a model

$$y(t)+a_1y(t-1)+\ldots+a_ny(t-n)=b_1u(t-1)+\ldots+$$

$$+b_mu(t-m)+e(t)+c_1e(t-1)+\ldots+c_ne(t-n) \tag{15}$$

Such a model is known as an ARMAX model.

<u>Output error models</u>. Instead of adding the disturbance $v(t)$ to the equation as in (14), it can be added as an output measurement error:

$$y(t) = x(t) + v(t) \tag{16a}$$

$$x(t)+f_1x(t-1)+\ldots+f_nx(t-n)=b_1u(t-1)+\ldots+b_mu(t-m) \tag{16b}$$

Such models are often called output error models. The "noise-free output" $x(t)$ is here not available for measurement, but given (16b) it can be reconstructed from the input. We denote by $x(t,\theta)$ the noisefree output that is constructed using the model parameters

$$\theta = (f_1\ldots f_n\ b_1\ \ldots b_m)^T, \tag{17}$$

i.e.

$$x(t,\theta)+f_1x(t-1,\theta)+\ldots+f_nx(t-n,\theta)=b_1u(t-1)+\ldots+b_mu(t-m) \tag{18}$$

With

$$\varphi(t,\theta)=(-x(t-1,\theta)\ldots-x(t-n,\theta)\ u(t-1)\ldots u(t-m))^T, \tag{19}$$

(16) can be rewritten as

$$y(t) = \theta^T\varphi(t,\theta) + v(t) \tag{20}$$

Notice the formal similarity to (14) but the important computational difference! Also in output error models the character of the additive noise v(t) can be further modelled.

State-space models. A common way of describing stochastic, dynamical systems is to use state-space models. Then the relationship between input and output is described by

$$x(t+1) = F(\theta)x(t) + G(\theta)u(t) + w(t)$$
$$y(t) = H(\theta)x(t) + e(t), \qquad (21)$$

where the noise sequences w and e are assumed to be independent at different time instants and have certain covariance matrices. Unknown, adjustable parameters θ may enter the matrix elements in F, G and H in an arbitrary manner. These may, e.g., correspond to canonical parametrizations (canonical forms) or to physical parameters in a time-continuous state space description that has been sampled to yield (21).

Models and Predictors

The list of potential models and model sets can be made long. For our purposes it is useful to extract the basic features of models, so as to allow for a treatment of model sets in general. First we introduce the follwing notation:

$\mathcal{M}(\theta)$: a particular model, corresponding to the parameter value θ

$\mathcal{M}$: a set of models:
$\mathcal{M}=\{\mathcal{M}(\theta)\,|\,\theta\in D_{\mathcal{M}} \subset R^d\}$

z^t: the set of measured input-output data up to time t:

$z^t=\{u(1),y(1),u(2),y(2)\ldots.u(t),y(t)\}$

Similarly, u^t and y^t denote the input sequence and the output sequence, respectively, up to time t.

The various models that can be used for dynamical systems all represent different ways of thinking and representing relationships between measured signals. They have one feature in common, though. They all provide a rule for computing the next output or a prediction (or "guess") of the next output, given previous observations. This rule is, at time t, a function from Z^{t-1} to the space where y(t) takes its values (R^p in general). It will also be parametrized in terms of the model parameter θ. We shall use the notation

$$\hat{y}(t|\theta) = g_{M}(\theta;t,Z^{t-1}) \tag{22}$$

for this mapping. The actual form of (22) will of course depend on the underlying model. For the linear difference equation (10)=(13) we will have

$$\hat{y}(t|\theta) = \theta^T\varphi(t) \tag{23}$$

The same prediction or guess of the output y(t) will be used for the model (14) with disturbances, in case $\{v(t)\}$ is considered as "unpredictable" (like white noise). For the state space model (21) the predictor function is given by the Kalman filter. Then g_M is a linear function of past data.

For the ARMAX-model (15) a natural predictor is computed as

$$\hat{y}(t|\theta)+c_1\hat{y}(t-1|\theta)+\ldots+c_n\hat{y}(t-n|\theta) =$$

$$= (c_1-a_1)y(t-1)+\ldots+(c_n-a_n)y(t-n)+b_1u(t-1) +$$

$$\ldots.+b_m u(t-m) \tag{24}$$

Notice that this can be rewritten as

$$\hat{y}(t|\theta) = \theta^T\varphi(t,\theta) \tag{25a}$$

$$\theta=(a_1\ldots.a_n\ b_1\ldots.b_m\ c_1\ldots.c_n)^T \tag{25b}$$

$$\varphi(t,\theta)=(-y(t-1)\ldots-y(t-n)u(t-1)\ldots u(t-m)\varepsilon(t-1,\theta)\ldots\varepsilon(t-n,\theta))^T \tag{25c}$$

$$\varepsilon(t,\theta) = y(t) - \hat{y}(t|\theta). \tag{25d}$$

For the model (16)=(20) a natural predictor is also given by (25a) with θ and $\varphi(t,\theta)$ defined by (7)-(19). Notice that in this case the prediction is formed from past inputs only. We then have, formally

$$\hat{y}(t|\theta) = g_M(\theta;t,u^{t-1}). \tag{26}$$

Such a model we call an "output error model" or a "simulation model".

Notice that the function $g_M(\theta;t,\cdot)$ in (22) is a deterministic function from the observations Z^{t-1} to the predicted output. All stochastic assumptions involved in the model descriptions (e.g. white noises, covariances matrices, Gaussianness) have only served as vehicles or "alibis" to arrive at the predictor function.

The prediction $\hat{y}(t|\theta)$ is computed from Z^{t-1} at time t-1. At time t the output y(t) is received. We can then evaluate how good the prediction was by computing

$$\varepsilon(t,\theta) = y(t) - \hat{y}(t|\theta). \tag{27}$$

We shall call $\varepsilon(t,\theta)$ the prediction error at time t, corresponding to model $M(\theta)$. This term will be the generic name for general model sets. Depending on the character of the particular model set, other names like, e.g. the (generalized) equation error may be used. For a simulation model (26) it is customary to call the corresponding prediction error (27) the output error.

We can also adjoin an assumption about the stochastic properties of the prediction error to the model $M(\theta)$:

"$M(\theta)$: Assume that the prediction error $\varepsilon(t,\theta)$ has the conditional (given Z^{t-1}) probability density function (p.d.f), $f(t,\theta,x)$ (28)

$$\left[\text{i.e. } P\left(\varepsilon(t,\theta)\in B\right) = \int_{x\in B} f(t,\theta,x)dx\right]"$$

Notice that in (28) there is an implied assumption of independence of the prediction errors, for different t, since the p.d.f does not depend on Z^{t-1}. A predictor model (22) adjoined with a probabilistic assumption (28) we shall call a probabilistic model.

All the linear models listed here can be summarized in the form

$$y(t)=G(q,\theta)u(t)+H(q,\theta)e(t) \tag{29}$$

with G and H as transfer functions of the shift operator q [$qu(t)=u(t+1)$]:

$$G(q,\theta)=\sum_{k=1}^{\infty} g_k(\theta)q^{-k}$$

$$H(q,\theta)=1+\sum_{k=1}^{\infty} h_k(\theta)q^{-k} \tag{30}$$

[The state space model (21) must then first be converted to innovations form].

The one step ahead predictor for (29) takes the form

$$\hat{y}(t|\theta)=H^{-1}(q,\theta)G(q,\theta)u(t)+[1-H^{-1}(q,\theta)]y(t) \tag{31}$$

5 FITTING PARAMETRIC MODELS TO DATA

Guiding Principles behind Identification Methods

The problem now, is to decide upon how to use the information contained in Z^N to select a proper member $\mathcal{M}(\hat{\theta}_N)$ in the model set, that is capable of "describing" the data. Formally speaking, we have to determine a mapping from Z^N to the set $\mathcal{M}$:

$$Z^N \rightarrow \mathcal{M}(\hat{\theta}_N) \tag{32}$$

Now, how can such mapping be determined? We pointed out that the essence of a model of a dynamical system is its prediction aspect. It is then natural to judge the performance of a given model $\mathcal{M}(\theta^*)$ by evaluating the prediction errors, $\varepsilon(t,\theta^*)$ given by (27). A guiding principle to form mappings (32), is thus the following one

"Based on Z^t, compute the prediction error $\varepsilon(t,\theta)$ using (22) and (27). At time $t=N$ select $\hat{\theta}_N$ so that the sequence of prediction errors $\varepsilon(t,\hat{\theta}_N)$, $t=1,\ldots,N$ becomes as small as possible".

The question is how to quantify what "small" should mean. Two approaches have been taken. These will be treated in the following two subsections.

Criterion minimization techniques

We introduce the scalar measure

$$\ell(t,\theta,\varepsilon(t,\theta)) \tag{33}$$

to evaluate "how large" the prediction error $\varepsilon(t,\theta)$ is. Here ℓ is a mapping from $R\times R^d\times R^p$ to R, where $d=\dim\theta$, $p=\dim y$.

After having recorded data up to time N a natural criterion of the validity of the model $\mathcal{M}(\theta)$ is

$$V_N(\theta,z^N) = \frac{1}{N}\sum_{1}^{N} \ell(t,\theta,\varepsilon(t,\theta)) \tag{34}$$

This function is, for given z^N, a well defined, scalar valued function of the model parameter θ. The estimate at time N, i.e., $\hat{\theta}_N$, is then determined by minimization of the function $V_N(\theta,z^N)$. This gives us a large family of well known methods. Particular, "named" methods are obtained as special cases, corresponding to specific choices of model sets and criterion functions $\ell(t,\theta,\varepsilon)$; and sometimes particular ways of minimizing (34).

The least squares method Choose $\ell(t,\theta,\varepsilon)=|\varepsilon|^2$ and apply the criterion (34) to the difference equation model (14). Since the prediction is given by (23) we have the prediction error

$$\varepsilon(t,\theta) = y(t) - \theta^T\varphi(t).$$

The criterion function (34) thus becomes

$$V_N(\theta,z^N) = \frac{1}{N}\sum_{1}^{N} |y(t)-\theta^T\varphi(t)|^2, \tag{35}$$

which we recognize as the familiar least squares criterion. See, e.g. Strejc (1980). This function is quadratic in θ, which is a consequence of the prediction being linear in θ and the quadratic choice of criterion function. This means that an explicit expression for the minimizing element $\hat{\theta}_N$ can be given:

$$\hat{\theta}_N = [\sum_1^N \varphi(t)\varphi^T(t)]^{-1} \sum_1^N \varphi(t)y^T(t). \tag{36}$$

□

A quadratic criterion like (36) is a common ad hoc choice also for general models. For multioutput systems this gives

$$\ell(t,\theta,\varepsilon) = \frac{1}{2}\,\varepsilon^T\Lambda^{-1}\varepsilon \tag{37}$$

To arrive at other specific functions ℓ, we could invoke, e.g. the maximum likelihood idea:

<u>The maximum likelihood method</u> For the probabilistic model (22), (28) the likelihood function can be determined. The calculations yield

$$\frac{1}{N}\log P(y(N),y(N-1),\ldots,y(1)) = \frac{1}{N}\sum_1^N \log f(t,\theta,\varepsilon(t,\theta)) \tag{38}$$

Maximizing the likelihood function is thus the same as minimizing the criterion (34) with

$$\ell(t,\theta,\varepsilon) = -\log f(t,\theta,\varepsilon) \tag{39}$$

For Gaussian prediction errors

$$-\log f(t,\theta,\varepsilon)=\text{const} + \frac{1}{2}\log\det\Lambda_t(\theta) + \frac{1}{2}\,\varepsilon^T\Lambda_t^{-1}(\theta)\varepsilon \tag{40}$$

where $\Lambda_t(\theta)$ is the assumed covariance matrix for the prediction errors. If the covariance matrix Λ_t is supposed to be known (independent of θ) then the first two terms of (40) do not affect the minimization, and we have obtained a quadratic criterion like (37). The maximum likelihood method was introduced for ARMAX models in Åström and Bohlin (1965).

□

For the least squares case it was possible to give an explicit expression for the parameter estimate. This is not the case in general. Then the criterion function (34) must be minimized using numerical search procedures. We shall comment more on this later.

We shall, following Ljung (1978), use the general term prediction error identification methods for the procedures we described in this section. See also Åström (1980). When applied to the special simulation model (26), the term output error methods might be preferred.

Correlation techniques

Another way of expressing that the sequence $\{\varepsilon(t,\theta)\}$ is small is to require that it be uncorrelated with a given sequence $\{\zeta(t)\}$. Let the vector $\zeta(t)$ represent some information that is available at time t-1.

$$\zeta(t) = \zeta(t,Z^{t-1}) \tag{41}$$

Sometimes, there is reason to consider a more sophisticated variant, where ζ itself may depend on the parameter θ. (Some such cases will be discussed below):

$$\zeta(t) = \zeta(t,\theta,Z^{t-1}) \tag{42}$$

The rationale for requiring $\varepsilon(t,\theta)$ and $\zeta(t)$ to be uncorrelated is the following: The predictors $\hat{y}(t|\theta)$ should ideally utilize all available information at time t-1. Thus, the prediction errors $\varepsilon(t,\theta)$ should be uncorrelated with such information. (If they are not, more information can be squeezed out from Z^{t-1}).

We thus determine $\hat{\theta}_N$ as the solution of

$$f_N(\theta,Z^N) \tag{43a}$$

where

$$f_N(\theta,Z^N) = \frac{1}{N}\sum_{t=1}^{N} \varepsilon(t,\theta)\zeta^T(t) \tag{43b}$$

where, normally, the dimension of ζ is such that (43) gives a system of equations that is compatible with the dimension of θ. When (43) is applied to the model (23) the well known instrumental variable method results. The vector ζ is then known as the instruments or the instrumental variables. See Söderström and Stoica (1983) for a further discussion of this method.

<u>How to choose $\zeta(t)$</u>? A way to make the estimate $\hat{\theta}$ insensitive to the characteristics of the noise that affects the system is to choose ζ to depend on past inputs only

$$\zeta(t) = \zeta(t,u^{t-1}) \tag{44}$$

Then that contribution to $\varepsilon(t,\theta)$ that has its origin in the noise will be uncorrelated with ζ for all θ. Choices (44) are typical for the instrumental variable method. It turns out that such choices that give the best accuracy of the obtained estimates are obtained when u is filtered through filters associated with the true system (see Söderström and Stoica, 1983). We then have

$$\zeta(t) = \zeta(t,\theta,u^{t-1}) \tag{45}$$

For models that can be written as (25a) (like the ARMAX model (15) and the output error model (16)=(20)) a natural choice is

$$\zeta(t,\theta,z^{t-1}) = \varphi(t,\theta) \tag{46}$$

Notice also that if we choose

$$\zeta(t,\theta,z^{t-1}) = -\phi(t,\theta)\Lambda^{-1} \tag{47}$$

where

$$\phi(t,\theta) = -\frac{d}{d\theta}\,\varepsilon(t,\theta) \tag{48}$$

we find that (43) will define the stationary points of the criterion (34), (37). The criterion minimization approach can thus be seen as a special case of (43), from this point of view.

<u>A frequency domain interpretation</u>

For the general linear model (29)-(31) we have that the prediction error is

$$\varepsilon(t,\theta)=H^{-1}(q,\theta)\left[y(t)-G(q,\theta)u(t)\right] \tag{49}$$

by Parsevals relationship we then have for $\ell(\varepsilon)=\varepsilon^2$

$$V_N(\theta,Z^N)=\frac{1}{N}\sum_{t=1}^{N}\varepsilon^2(t,\theta)\approx\int_{-\pi}^{\pi}|Y_N(\omega)-G(e^{i\omega},\theta)U_N(\omega)|^2/|H(e^{i\omega},\theta)|^2d\omega=$$

$$=\int_{-\pi}^{\pi}|\hat{G}_N(e^{i\omega})-G(e^{i\omega},\theta)|^2|U_N(\theta)|^2/|H(e^{i\omega},\theta)|^2d\omega \qquad (50)$$

Here Y_N and U_N are the DFT:s of y^N and u^N and $\hat{G}_N$ is defined by (8).

Minimizing $V_N(\theta,Z^N)$ can thus be interpreted as fitting $G(e^{i\omega},\theta)$ to the Fourier transform estimate $\hat{G}_N(e^{i\omega})$ in a quadratic norm (50) with the "signal-to-noise ratio" $|U_N(\omega)|^2/|H(e^{i\omega},\theta)|^2$ as weighting function. This shows a conceptual relationship between the prediction error estimates and the spectral analysis estimate (9).

Computing the estimate

By

$$\hat{\theta}_N=\arg\min V_N(\theta,Z^N) \qquad (51)$$

or

$$\hat{\theta}_N=[\text{sol } f_N(\theta,Z^N)=0] \qquad (52)$$

["arg min" V: the minimizing argument of V "sol f=0": the solution to the equation f=0], we have (implicitly) defined the estimate. It now "only" remains to compute it. This could be a laborous procedure, but is as such a standard problem in numerical analysis. See, e.g., Dennis and Schnabel (1983) for a recent treatment. In addition, the system identification field is characterized by a considerable amount of different algorithms, many of which can be interpreted as attempts to solve (51) or (52), perhaps utilizing specific structures that stem from the particular prediction function (22) employed. See Ljung (1982) for some aspects on this.

An important class of special minimization and solution methods is formed by procedures that are required to work sequentially with the data stream; recursive identification methods. See e.g. Ljung and Söderström (1983) for further discussions of this topic.

6 ASYMPTOTIC PROPERTIES OF IDENTIFIED MODELS

Analysis

In the previous chapter we listed general principles, as well as particular methods, for system identification. We noted that an identification method generally can be seen as a mapping (32) from the observed data Z^N to a particular member in the model set $\mathcal{M}$.

We now ask the question: What are the properties of this mapping and of the resulting model? This question can be answered basically in two ways:

1. Generate data Z^N with known characteristics. Apply the mapping (corresponding to a particular identification method) and evaluate the properties of $\mathcal{M}(\hat{\theta}_N)$. This is known as simulation studies.

2. Assume certain properties of of Z^N and try to calculate what the inherited properties of $\mathcal{M}(\hat{\theta}_N)$ are. This is known as analysis.

The mapping (32) is in general fairly complicated and non-linear. Moreover, the data Z^N are often considered to be a realization of a stochastic process. This means that although the mapping itself is deterministic the analysis has to be performed in a probabilistic framework. The parameter $\hat{\theta}_N$ will be a random variable with a distribution that is inherited from the probabilistic properties that we in general can describe only the asymptotic properties of $\hat{\theta}_N$ as N tends to infinity. The typical asymptotic aspects are to establish to what $\hat{\theta}_N$ converges as $N \to \infty$ and what the asymptotic dsitribution of $\hat{\theta}_N$ is.

In this section we shall discuss such asymptotic, analytic results. The analysis is a fairly technical problem, and it is beyond the scope of these notes to provide the theory itself. Instead, we shall quote the bottom line of the analysis, which has direct implications for the user choices to be discussed in section 7. Furthermore, we shall here confine ourselves to the criterion minimization approach.

The fact that the analysis only deals with asymptotic results implies certain limitations. With asymptotic results we know the properties of $\hat{\theta}_N$ for "large" N. However the theory usually does not provide any information about how large N has to be for the results to be applicable. It may be $N \approx 10^6$, which obviously makes a big difference to the user. Therefore, to get some insight into the properties of $\hat{\theta}_N$ for realistic values of N, the analysis must be complemented with simulation studies.

Convergence

Consider the criterion (34)

$$V_N(\theta, Z^N) = \frac{1}{N} \sum_{1}^{N} \ell(t, \theta, \varepsilon(t, \theta)). \tag{53}$$

The minimizing value of θ is denoted by $\hat{\theta}_N$.

Remark. By $\hat{\theta}_N$ we understand the value that yields to global minimum of (53), or if the minimum is not unique, one of the values that minimize it. This $\hat{\theta}_N$ may of course differ from the value the user actually obtained when numerically minimizing (53) depending on how the numerical serach for a minimum is performed.

□

In this section, we shall without proofs quote some results from Ljung (1978) and Ljung and Caines (1979).

Suppose that the limit

$$\bar{V}(\theta) = \lim_{N \to \infty} E\, V_N(\theta, Z^N) \tag{54}$$

exists, where "E" means expectation with respect to Z^N. Then, under weak regularity conditions,

$\hat{\theta}_N$ converges w.p.1 to a minimum of $\bar{V}(\theta)$ as N tends to infinity. (55)

This is true, whether or not the model set M is capable of a true description of the data. It is thus quite a general result.

Let us now specialize to the following assumption

- $\ell(t,\theta,\varepsilon)=\varepsilon^2$ (56a)

- The model set (29) is used (56b)

- The data Z^N has been generated by "the true system"

$$S\colon\ y(t)=G_0(q)u(t)+H_0(q)e(t) \qquad (56c)$$

  where $\{e(t)\}$ is a sequence of independent random variables with zero mean values and variance λ.

- Introduce

$$D_T(S,M)=\{\theta\,|\,G(e^{i\omega},\theta)=G_0(e^{i\omega});H(e^{i\omega},\theta)=H_0(e^{i\omega}) \text{ for all } \omega\} \qquad (56d)$$

  and assume that $D_T(S,M)$ is non-empty.

- $e(t)$ is independent of y^{t-1} and u^{t-1} (56f)

It is now easy to verify that under (56) we have

$$\hat{\theta}_N \to D_T(S,M) \text{ w.p.1 as } N\to\infty \qquad (57)$$

which means that the estimates of the transfer functions are strongly consistent

Asymptotic distribution

Suppose now that $\hat{\theta}_N$ converges of θ^*, such that the matrix

$$\frac{d^2}{d\theta^2}\,\bar{V}(\theta^*)$$

is invertible. Then

$$\sqrt{N}\,(\hat{\theta}_N-\theta^*) \in \text{As } N(0,P) \qquad (58)$$

which means that the random variable $\sqrt{N}\,(\hat{\theta}_N-\theta^*)$ converges in distribution to the normal distribution with zero mean and covariance matrix P. Here

$$P=[\bar{V}''(\theta^*)]^{-1} \quad \lim_{N\to\infty} N\cdot E\; V_N'(\theta^*)[V'_N(\theta^*)]^T \quad [\bar{V}''(\theta^*)]^{-1} \tag{59}$$

where prime and doubleprime, denotes differentation once and twice, respectively, w.r.t θ. If θ^* is a value such that the assumption (28) is true for $\theta=\theta^*$, then it can be shown that the matrix P in (59) equals the Cramér-Rao lower bound for the covariance matrix of any (unbiased) estimator.

Let us evaluate the expression (59) under assumptions (56)

$$P=[E\phi(t,\theta_0)\phi^T(t,\theta_0)]^{-1}\phi(t,\theta_0)\lambda\phi^T(t,\theta_0)\times$$

$$\times[E\phi(t,\theta_0)\phi^T(t,\theta_0)]^{-1}=\lambda[E\phi(t,\theta_0)\phi^T(t,\theta_0)]^{-1} \tag{60}$$

where $\phi(t,\theta)$ is the gradient of the prediction

$$\phi(t,\theta) = \frac{d}{d\theta}\,\hat{y}(t|\theta) \text{ (column vector of dimension = dim}\theta) \tag{61}$$

and θ_0 is a value in $D_T(S,M)$.

This result has an interesting implication. Let

$$\bar{V}(\theta) = E\;\varepsilon^2(t,\theta) \tag{62}$$

Then we can evaluate how good the model $M(\hat{\theta}_N)$ is by calculating the value

$$\bar{V}(\hat{\theta}_N).$$

Here $\hat{\theta}_N$ is a random variable, and we may evaluate the expectation of $\bar{V}(\hat{\theta}_N)$ w.r.t $\hat{\theta}_N$. This gives, after some straightforward calculations

$$E\bar{V}(\hat{\theta}_N)\sim E\;\bar{V}(\theta_0)+\lambda\frac{\dim\theta}{N}, \tag{63}$$

using (60). Here "~" means asymptotically equal to.

The result (63) is remarkable in its generality. It tells us that the expected prediction error variance increases with the number of independent parameters in the model (once the model set is large enough to make $D_T(S,M)$ non-empty) irrespectively of where the parameters enter the model.

Frequency domain expressions

So far we have focused our attention on the parameter vector θ. In many cases black-box models are employed and then the parameters are just means to describe a flexible set of models. For linear systems this means that we are more interested in the transfer function estimates

$$\hat{G}_N(e^{i\omega})=G(e^{i\omega},\hat{\theta}_N)$$
$$\tag{64}$$
$$\hat{H}_N(e^{i\omega})=H(e^{i\omega},\hat{\theta}_N)$$

than in $\hat{\theta}_N$ itself. It is therefore useful to characterize the properties of (64) directly.

First, with (56abc) we have

$$\bar{V}(\theta)=E\varepsilon^2(t,\theta)=\frac{1}{2\pi}\int_{-\pi}^{\pi} E\left|Y(\omega)-G(e^{i\omega},\theta)U(\omega)\right|^2/\left|H(e^{i\omega},\theta)\right|^2 d\omega=$$

$$=\frac{1}{2\pi}\int_{-\pi}^{\pi}\left[\left|G_0(e^{i\omega})-G(e^{i\omega},\theta)\right|^2\Phi_u(\omega)+\lambda\left|H_0(e^{i\omega})\right|^2\right]/\left|H(e^{i\omega},\theta)\right|^2 d\omega \tag{65}$$

provided $\{u(t)\}$ and $\{e(t)$ are independent. Here $\Phi_u(\omega)$ is the input spectrum.

The expression (65) together with (55) now characterizes the limiting estimate. If G and H are independently parameterized:

$$G(q,\theta)=G(q,\rho); \quad H(q,\theta)=H(q,\eta), \quad \theta=\begin{pmatrix}\rho\\ \eta\end{pmatrix} \tag{66}$$

and

$$\begin{pmatrix}\rho^*\\ \eta^*\end{pmatrix}=\theta^*=\arg\min \bar{V}(\theta) \tag{67}$$

then (67) can be rewritten

$$\rho^* = \arg\min_{\rho} \int_{-\pi}^{\pi} |G_0(e^{i\omega}) - G(e^{i\omega},\rho)|^2 \Phi_u(\omega)/|H(e^{i\omega},\eta^*)|^2 d\omega \tag{68}$$

$$\eta^* = \arg_{\eta}\min \int_{-\pi}^{\pi} [\Phi_{ER}(\omega)]/|H(e^{i\omega},\eta)|^2 d\omega$$

$$\Phi_{ER}(\omega) = |G_0(e^{i\omega}) - G(e^{i\omega},\rho^*)|^2 \Phi_u(\omega) + \lambda |H_0(e^{i\omega})|^2 \tag{69}$$

This shows that $\hat{G}$ is fitted to G_0 in a quadratic frequency-domain norm using the weighting function $\Phi_u(\omega)/|H(e^{i\omega},\eta^*)|^2$ (the model's signal-to-noise ratio). Moreover the model spectrum $|\hat{H}(e^{i\omega})|^2$ is fitted to the error spectrum (69). More consequences of the expression (65) are discussed in Wahlberg and Ljung (1984).

For the asymptotic distribution the following rather remarkable result can be shown: Consider the black-box model set (29) being of order n. Let the estimates be $\hat{G}_N(e^{i\omega},n)$ and $\hat{H}_N(e^{i\omega},n)$ and let the limit values as $N\to\infty$ be $G^*(e^{i\omega},n)$ and $H^*(e^{i\omega},n)$. Then

$$\sqrt{N}\begin{pmatrix} \hat{G}_N(e^{i\omega},n) - G^*(e^{i\omega},n) \\ \hat{H}_N(e^{i\omega},n) - H^*(e^{i\omega},n) \end{pmatrix} \in AsN\big(0, P_n(\omega)\big) \tag{70}$$

$$\lim_{n\to\infty} \frac{1}{n} P_n(\omega) = \Phi_v(\omega) \begin{bmatrix} \Phi_u(\omega) & \Phi_{ue}(\omega) \\ \Phi_{eu}(\omega) & \lambda \end{bmatrix}^{-1} \tag{71}$$

Here $\Phi_{ue}(\omega)$ is the cross spectrum between input $\{u(t)\}$ and noise $\{e(t)\}$ (which is zero if the system operates in open loop). $\Phi_v(\omega)$ is the additive noise spectrum: $\Phi_v(\omega) = \lambda|H_0(e^{i\omega})|^2$.

For a formal statement and proof of this result, see Ljung (1984b).

If the identification experiment has been performed in open loop we thus have for the asymptotic variance of $\hat{G}$:

$$\operatorname{Var} \hat{G}_N(e^{i\omega},n) \sim \frac{n}{N} \cdot \frac{\Phi_v(\omega)}{\Phi_u(\omega)} \tag{72}$$

7. Some User Choices

To determine a good identification procedure for a given situation means that the user has to make a number of choices. He or she has to select experimental conditions (which signals to measure, which inputs to use, sampling rates etc), the set of models, and the functions ℓ in (33) or ζ in (41). The asymptotic results quoted here, like the set D_c, into which the estimates converge (see (55)) and the asymptotic covariance matrix P (see (58) -(60)), are indeed functions of these listed design variables. Rational decisions of the design variables can thus be based on analysis of D_c and P. This analysis may be complex, but certain conclusions can be drawn from general considerations:

- o The experimental conditions should be such that the predictions become sensitive w.r.t interesting parameters (see (60)): If ψ is "large" then P will be "small".

- o The model set should be parsimoneous when describing the system (use few parameters) according to (31);

- o The choice of ℓ should be matched to the probabilistic properties of the prediction errors (optimal choice $\ell(\varepsilon)=-\log f(x)$, where $f(x)$ is the probability density function of $\varepsilon(t,\theta_0)$).

- o Since, the best approximation of the system for the experimental conditions at hand is obtained asymptotically, the chosen conditions should resemble those for which the model is to be used.

In addition to this general advice, the asymptotic expressions (70)-(72) can be used to calculate some fairly explicit, asymptotic optimal, experiment designs. We examplify this for a simple special case:

Suppose that the design objective is to minimize a criterion

$$\int_{-\pi}^{\pi} \text{Var}\, \hat{G}_N(e^{i\omega}) \cdot C(\omega)\, d\omega \tag{73}$$

with respect to the input spectrum subject to constrained input variance. With (72) this problem takes the following mathematical form:

$$\min_{\Phi_u(\cdot)} \int_{-\pi}^{\pi} \frac{\Phi_v(\omega)}{\Phi_u(\omega)} \cdot C(\omega)d\omega \tag{74}$$

$$\text{subject to} \int_{-\pi}^{\pi} \Phi_u(\omega)d\omega \leq \alpha$$

This problem has the solution

$$\Phi_u^{opt}(\omega) = \mu \cdot \sqrt{\Phi_v(\omega) \cdot C(\omega)} \tag{75}$$

where μ is a constant, adjusted so that equality is obtained in the constraint. See Ljung (1984b) for more details.

8. MODEL VALIDATION

The model validation problem is to decide whether the model obtained from the identification procedure can be accepted. (It could also extend to the situation where we would like to check the validity of a model that has been obtained in some other way). If the identification method has provided the best possible model in the model set (as it is supposed to do) we might say that the problem is to decide whether the chosen model set M was a suitable one. The procedure must therefore be linked with the aspects on the model set that were discussed in Section 4.

The choice of model set involves three different decisions:

- o What "type" of model set should be used (e.g. linear or non linear, a difference equation or an ARMAX model, etc) (76)

- o What "size" should be chosen (e.g. order of a state space model) (77)

- o What "parametric structure" should be used (a state space model of given order can be parametrized in different, non-equivalent ways) (78)

These decisions thus become part of the model validation procedure.

The question "Is the identified model good enough?" can be approached in two ways. One is to evaluate the properties of the model and decide whether it meets reasonable requirements. Another idea is to shop around in other model sets and compare with the models offered there. The latter procedure can be regarded as an a posteriori (i.e. after the data has been processed) choice of model set. It thus offers a way to make the selections (76) - (78) with the aid of data. These choices can also be made using direct methods based on data analysis of a different character than the identification procedure. It may be questionable whether such methods should be considered as "model validation", but refer to Unbehauen and Göring (1974) and van den Boom and van den Enden (1974).

<u>A Posteriori comparison of different model sets</u>

A natural idea to solve the model set choice is to test different model sets and decide which one to use on the basis of how they perform. "Performance" can of course be evaluated in various ways, but the dominating one is to use the criterion itself, just as for the evaluation within the model set. Suppose that an identification criterion $V_N(\theta,Z^N)$ (see (34)) is used.

Supposing the minimising value in model set

$\mathcal{M}_i$ is $\hat{\theta}_N^{(i)}$, we can evaluate

$$W_N(\mathcal{M}_i,Z^N):=V_N(\hat{\theta}_N^{(i)},Z^N), \tag{79}$$

and base the comparison of different $\mathcal{M}_i$:s on W. Now, this is not entirely straightforward. A larger model set $\mathcal{M}_i$ $\mathcal{M}_k$ will always give a lower value of (79). Therefore we must take some prejudice against more complex model sets.

A natural idea to introduce the prejudice against complex model sets, mentioned at the end of the previous subsection, is to introduce a penalty term:

$$W_N(\theta,\mathcal{M},Z^N) = V_N(\theta,Z^N)(1+U_N(\mathcal{M})) \tag{80a}$$

or

$$W_N(\theta, M, Z^N) = V_N(\theta, Z^N) + U_N(M) \tag{80b}$$

Here V is the criterion function and $U_N(M)$ is some measure of complexity of the model set M.

The use of (80) and the choice of $U_N(M)$ can be approached in a pragmatic fashion:"If I am going to accept a more complex model (according to my own complexity measure) it has to prove to be significantly better". The criterion function (80) can however also be approached in a formal way. The original contribution of this character is Akaike's final prediction error (FPE) criterion. The question posed then is:"Suppose I use the obtained model for prediction on another data set. Which model(set) then gives the smallest prediction error (variance)?". The answer is that (80a) should be minimized with respect to M and θ for

$$U_N(M) = 2 \cdot \frac{\dim\theta}{N}. \tag{81}$$

Akaike's original treatment, Akaike (1969) was for the quadratic case $\ell(t,\theta,\varepsilon)=\frac{1}{2}\ \varepsilon^T\Lambda\varepsilon$. He has later, e.g., Akaike (1974) extended the idea to a general situation based on information theoretic considerations. The resulting criterion, AIC, is still given by (80a), (34) and ℓ given by the maximum likelihood choice (see eq (39).

The selection between two model sets M_0 and M_1 can also be approached by the theory of statistical tests. The idea is to pose the hypothesis

$$H_0\text{: The data has been generated by } M_0(\hat{\theta}_N^{(0)}) \tag{82}$$

This hypothesis is to be tested against the alternative

$$H_1\text{; The data has been generated by } M_1(\hat{\theta}_N^{(1)}). \tag{83}$$

To introduce a prejudice against complex models now means that we would prefer the set M_0 unless there is "convincing evidence" that the hypothese H_1 is true. In statistics this is expressed as that H_0 is the "null hypothesis".

A closer analysis, see Söderström (1973), shows that such hypothesis tests are closely related to criteria like (80) from an operational view.

Can the obtained model be accepted?

We have so far discussed how to compare different model sets. We have, hopefully, settled for a model that is not only the best one in the chosen model set, but also compares favourably to models in other model sets. But the true problem of model validation has not been addressed:"The obtained model may be the best one available within reasonable model sets. But is it good enough?".

This question can be interpreted in two ways. "Is the model good enough with respect to my purpose?" or "Is the model good enough with respect to the data record?".

Model validation with respect to the purpose of the modelling: There is always a certain purpose with the modelling. It might be that the model is required for regulator design, for prediction, for simulation etc. The ultimate validation then is to test whether the problem that motivated the modelling exercise can be solved using the obtained model.

If a regulator based on the model gives satisfactory control, then the model was a "valid" one, regardless of the formal aspects of this concept that can be raised.

A commonly applied procedure that can be regarded as a test of the model's validity for simulation is to simulate the system with the actual input and compare measured output with the simulated model output. Prefer able a different data set should be used for this comparison, then was used for the estimation of $\hat{\theta}_N$.

Model validation with respect to measured data: After the identification procedure we are left with the data set y^N, u^N and with a model $\mathcal{M}(\theta_N)$. The general model can be interpreted as a probabilistic model for the generation of y; see (22), (28).

The model validation question, related to the data, then asked is:

"What is the probability that the data record y^N, u^N actually has been generated by the model (22), (28) for $\theta=\hat{\theta}_N$?"

(Well, the question has to be somewhat modified, since any probability of this kind is zero)

This question is equivalent to:

"What is the probability that the sequence

$$\varepsilon(t,\hat{\theta}_N):=y(t)-g_M(\theta_N,t,z^{t-1})=y(t)-\hat{y}(t|\hat{\theta}_N)$$

is a realization of a sequence of independent random vectors with p.d.f $f(\hat{\theta}_N,t,x)$?"

For such a problem several different tests (whiteness tests) have been developed, see, e.g. Jenkins and Watts (1968). A typical variant is to evaluate

$$r_\varepsilon(k)=\frac{1}{N}\sum_{t=1}^{N-k}\varepsilon(t,\theta_N)\varepsilon^T(t+k,\theta_N)\quad k=1,2,\ldots,M$$

and test the size of these numbers, using e.g. the quantity

$$\xi_{N,M}=\sum_{k=1}^{M}|r_\varepsilon(k)|^2.$$

9 CONCLUSIONS AND OUTLOOK

System identification has been a central research area in the control field over three decades. It has been characterized by quite diverse approaches and a multitude of suggested "methods". Potential users may find the area incoherent when consulting the literature. Recent years have brought more unification into the field and it may be expected that the different views on identification techniques are converging to a more coherent picture.

It is my feeling that the System Identification has reached a certain state of maturity: Several methods (packaged into nice interactive soft-ware products) have become everyday tools of analysis in many application areas, the asymptotic theory is fairly well understood and efficient algorithms are available. No doubt, several issues remain to be explored but it does not seem likely that they will overthrow the foundations outlined in this contribution.

One might hope that the insight and knowledge of statisticians would be more fully utilized and adapted to the system identification problem. This concerns in particular how to support the important choice of model set by inference from data. Good techniques are established for the model order choice in black-box type model sets, but insights for more difficult choices of model sets would be desirable.

On the other hand, the control-oriented identification literature has traditionally been more inclined to use advanced results and models, than, say, the signal processing literature. This is of course due to the latter's focus on simple and fast algorithms, resulting from rapid data streams. With the computer revolution one may expect an increased interest in sophisticated signal processing algorithms and perhaps a transfer of identification methods from the control area.

Identification, adaptation and robustnesss

Has adaptive and robust control made identification obsolete? Before answering this question, let us formalize the design problem as

$$\min_{\rho} J(\rho, S) \tag{90}$$

Here ρ are the design (regulator) parameters and S denotes the characteristics of the known system. The optimal regulator will be

$$\rho(S) \tag{91}$$

The source of uncertainty is of course that S is not known. The "identification approach" to tackle the uncertainty is to determine a model M_N approximating S and using the design

$$\rho(\hat{M}_N) \tag{92}$$

Adaptation can, loosly, be described as realizing (90) be on-line adjustment of ρ. However most successful adaptation methods (cf Åström (1983)) indeed achieve this by estimating $\hat{M}_N$ recursively and applying (92) with the current estimate. Identification is thus a key ingredient in adaptation mechanisms!

A possible problem with these solutions is that $H(\rho, S)$ may be sensitive with respect to S at $\rho=\rho(S)$. Then $\hat{M}_N$ may perhaps never become good enough an approximation of S. The robustness approach, again loosely, is then to replace (90) by

$$\min_{\rho} \max_{S \in \mathbf{S}} J(\rho, S) \tag{93}$$

thereby reducing the uncertainty form S to the set S. However, for (93) to yield good performance, and the design to be not too conservative, it is desirable to let the set S be "small". This is a goal that can be achieved by identification. The results (58)-(60) and (70)-(71) indeed provide us with such sets. Regardless of which approach we use to handle uncertainty, identification is thus not obsolete!

REFERENCES

H. Akaike. Fitting autoregressive models for prediction, Ann Inst Statis Math Vol 21, 1969, pp 243-247.

H. Akaike. A new look at the statistical model identification, IEEE Trans Automatic Control, Vol AC-19, 1974, pp 716-723.

K J Åström (1980). Maximum likelihood and prediction error methods. Automatica, Vol 16, pp 551-574.

K. J. Åström (1983). Theory and applications of adaptive control - a survey, Automatica , Vol 19, pp 471-486.

K. J. Åström and T. Bohlin (1965). Numerical identification of linear systems from normal operating records, IFAC Syposium on self-adaptive systems, Teddington, England. Also in P.H. Hammond, ed.: Theory of self-adaptive control systems, Plenum Press, New York.

K.J. Åström and P. Eykhoff (1971). System identification - a survey. Automatica, Vol 7, pp 123-167.

A J W van den Boom and A W M van den Enden (1974). The determination of the order of process and noise dynamics. Automatica, Vol 10, pp 245-258.

D.R. Brillinger (1981). Time series-Data Analysis and Theory, Holden-Day, San Francisco.

J.E. Dennis and R. B. Schnabel (1983). Numerical Methods for Unconstrained Optimization and Non-linear Equations, Prentice-Hall, Englewood Cliffs.

K. H. Fasol and H. P. Jörgl (1980). Principles of Model Building and Identification, Automatica, vol 16, pp 505-518.

M. Gevers and G. Bastin (1982). What does system identification have to offer? Proc IFAC Symp on Identification, Washington DC, pp 29-36.

K. R. Godfrey (1980). Correlation methods, Automatica, vol 16, pp 527-534.

G.C. Goodwin and R. L. Payne (1977). Dynamic System Identification: Experiment Design and Data Analysis, Academic Press.

N.K. Gupta and R.K. Mehra (1974). Computational aspects of maximum likelihood estimation and reduction in sensitivity function calculations, IEEE Trans. Autom. Control. Vol AC-19, pp 744-783.

G.M. Jenkins and D.G. Watts (1969). Spectral Analysis and its Applications, Holden-Day, San Fransisco.

H. Kwakernaak, H and R. Sivan (1972). Linear Optimal Control Systems, Wiley, New York.

L. Ljung (1978). Convergence analysis of parametric identification methods. IEEE Trans on Autoamtic Control, Vol Ac-23, pp 770-783.

L. Ljung (1982). Identification methods, Proc IFAC Symp on Identification, Washington DC, pp 11-18.

L. Ljung (1984). On the Estimation of Transfer Functions, Report LiTH-ISY-I-0692, Department of Electrical Engineering, Linköping University.

L. Ljung (1984b) Asymptotic variance expressions for identified black-box transfer function models, 23rd IEEE CDC, Las Vegas, pp 951-958.

L. Ljung and P. E. Caines (1979). Asymptotic normality of prediction error estimation for approximate system models, Stochastics, Vol 3, pp 29-46.

L. Ljung and T. Söderström (1983). Theory and Practice of Recursive Identification, MIT Press, Cambridge, Mass.

H. Rake (1980). Step response and frequency response methods. Automatica, Vol 16, pp. 519-526.

T. Söderström (1977). On model structure testing in system identification, Int J Control, Vol 26, pp 1-18.

T. Söderström and P. G. Stoica (1983). Instrumental Variable Methods for System Identification, Springer-Verlag, Berlin.

V. Strejc (1980). Least squares parameter estimtion. Automatica, Vol 16, pp 535-550.

H. Unbehauen and B. Göhring (1974). Tests for determining model order in parameter estimation, Automatica, Vol 10, pp 233-244.

B. Wahlberg and L. Ljung (1984) Design variables for bias distribution in transfer function estimation, 23rd IEEE CDC, Las Vegas, pp 335-341.

P.E. Wellstead (1979). Introduction to Physical System Modelling, Academic Press, London.

J. Wieslander (1979). Interaction in computer aided analysis and design of control systems. Thesis CODEN: LUF 122/(TFRT-1019)/1-222. Dept of Automatic Cotnrol, Lund Institute of Technology, Lund, Sweden.

UNCERTAINTY MODELS AND THE DESIGN OF ROBUST CONTROL SYSTEMS

Huibert Kwakernaak
Twente University of Technology
7500 AE Enschede, The Netherlands

Abstract

Several models to represent uncertainty in control systems are reviewed, and a survey is given of various methods for the design of insensitive and robust control systems.

Contents

1. Introduction

Control science is concerned with the problem how to guide and regulate systems such that they exhibit a certain desired behavior. To a large extent, the difficulties that arise in realizing this goal stem from disturbances that affect the system and uncertainty about the system behavior. It was found a long time ago that feedback is a very effective way to cope with disturbances and uncertainties.

Feedback theory underwent a strong development during the 1930's. This growth was brought about by new technical applications made possible by the development of the vacuum tube. These developments were absorbed into control theory during the 1940's. During the same period, Wiener did his work on optimal filtering. Wiener's work had two aspects whose impact on control theory cannot be underestimated:

the use of stochastic models (in particular stochastic processes) to represent uncertainty, and the idea to translate design problems (in his case that of finding a filter) into optimization problems.

Wiener's work was assimilated into control theory during the 1950's, with as culmination perhaps the well-known book by Newton, Gould and Kaiser (1957). Beginning in this same period, the idea of solving design problems by way of optimization greatly took the fancy of control theorists. This led to the optimal control era, which can be dated to the 1960's. Unfortunately, during this period, "control" was interpreted in the sense of "steering", and the paramount importance of feedback was at times lost from view. Feedback emerged more or less incidentally as a convenient property of solutions of optimal control problems formulated in a state space framework, particularly when approached from the point of view of dynamic programming. In stochastic control theory it fortunately was recognized that feedback is an undeniable aspect of optimal solutions under uncertainty.

The late 1960's formed the beginning of the rise of "LQG"-theory. This approach relies on stochastic optimal control theory and the solutions it presents therefore involve feedback. Although LQG has enriched control theory in many respects, it also has some shortcomings. LQG can deal admirably with disturbances but only indirectly with other forms of uncertainty.

LQG and its refinement geometric control theory, the mathematical consolidation of stochastic optimal control theory, and subjects like optimal filtering, hierarchical and decentralized control, realization theory and identification took control theory through the 1960's and much of the 1970's. Coping with uncertainty has always remained a problem of considerable interest. One proof of this is the continuing work on adaptive systems, which is currently attracting a revitalized attention. Also during the last few years there has been renewed emphasis on exploiting feedback to cope with uncertainty. Some of this work will be reviewed later.

The purpose of the present contribution is first of all to review a number of paradigms in control theory to represent uncertainty. In Section 2 we shall distinguish between disturbances and perturbations. In Section 3 this distinction leads to a discussion of the notions of sensitivity and robustness. Section 4 surveys some methods for the design of insensitive control systems, while Section 5 deals with the design of robust control systems. The conclusions are summarized in Section 6.

2. Disturbance and perturbation

In the discussion of control systems, usually a distinction is made between two kinds of adverse effects: disturbances, and perturbations. To make this distinction clear, the situation is made more concrete. The basic situation is represented by the block diagram of Fig. 1. Given is a dynamic system with input u and controlled output z. This output is required to exhibit a certain behavior: in the case of a regulating system z should be constant and equal to a given value, in the case of a servo-type system z should follow a given reference signal. Then a disturbance is an external signal that influences the plant and whose behavior cannot be controlled (Fig. 2). A perturbation on the other hand is a transient or permanent change in the dynamic and/or static properties of the plant (Fig. 3).

The distinction is not always evident. Perturbations have the effect of inducing disturbances by parametric modulation of the regular input u. This can be clarified as follows. Let the plant be described as

$$z = Hu, \tag{1}$$

where H is some operator (linear or nonlinear, time-invariant or time-varying. Then if the unperturbed ("nominal") plant is represented by

$$z_o = H_o u \tag{2}$$

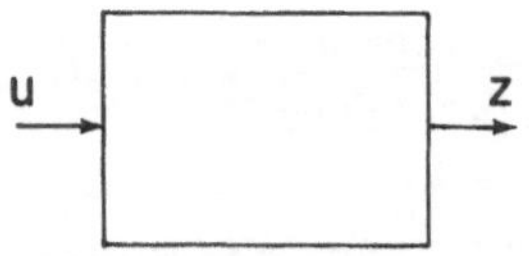

Fig. 1. Plant

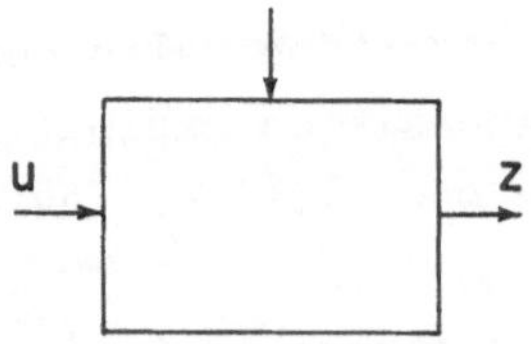

Fig. 2. Plant with disturbance

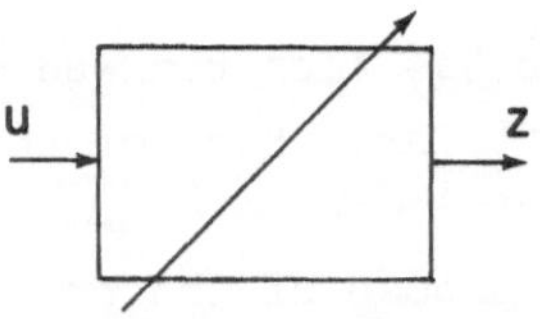

Fig. 3. Perturbed plant

for the same input u, then we can write

$$z = Hu = H_o u + v \tag{3}$$

where

$$v := Hu - H_o u \tag{4}$$

can be seen as an output-additive disturbance (Fig. 4). It is clear, however, that this induced disturbance is strongly coupled with the plant input u. Therefore perturbations cannot always be treated as disturbances although this is sometimes done.

We shall now discuss various more detailed possible assumptions about the disturbances and perturbations. With regard to the disturbances it sometimes is known at which specific points in the plant the disturbance enters, so that the plant equations may accordingly be modified. This leads in the case of finite-dimensional linear state space models to equations of the form $\dot{x} = Ax + Bu + Gv$, where v is the disturbance, and the coefficient matrix G specifies the points of entry of the disturbance. In other situations for lack of detailed information the effect of the disturbances is collected at the plant output (Fig. 5).

For the nature of the disturbances any of the following can be envisaged:

(a) The time behavior of the disturbance is unknown, and any shape is possible. In this situation it is usually assumed that the signal is subject to certain limitations, e.g. that it has bounded amplitude or bounded energy.

(b) The disturbance has a known deterministic shape with certain unknown parameters. (step with unknown starting time and height;

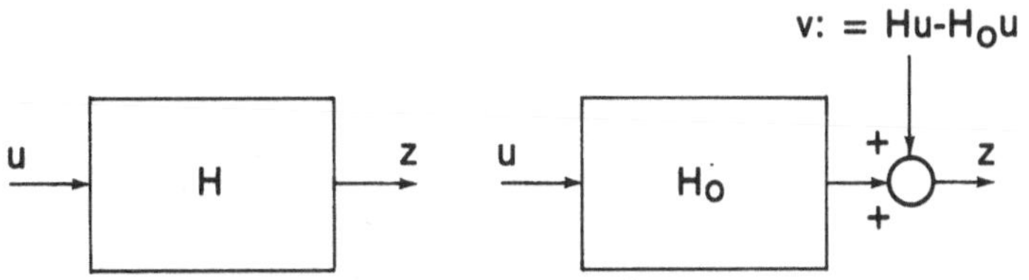

Fig. 4. Perturbed plant (left) represented as unperturbed plant with induced disturbance (right).

sinusoid with unknown frequency, amplitude and phase, etc.). In this situation usually some bounds on the parameters are assumed.

(c) The disturbance is shape-deterministic as in (b), but the parameters are random variables with a known joint distribution.

(d) The disturbance is a realization of a random process with known stochastic properties (e.g. white noise or exponentially correlated noise).

(e) The disturbance is a realization of a random process with partially known stochastic properties (e.g. Gaussian noise with unknown power spectral density). Also here usually some bound is assumed, e.g. that the process has finite power.

For practical engineering purposes, such as control system performance specification and evaluation, usually shape-deterministic disturbances are assumed. For theoretical work any of the other models is used as appropriate.

In the representation of perturbations any of the following situations may arise.

(A) A complete plant model is available but there is uncertainty about the numerical values of some of the parameters. This uncertainty may be represented as follows:

(A.1) the parameter values are unknown but bounds on their ranges of variation are specified;

(A.2) the parameters are random variables with a completely or partially known joint distribution.

(B) A plant model is available (possibly with some uncertain parameters) but it is known that in setting up the model certain marginal effects that nevertheless affect the control system performance have been omitted or neglected.

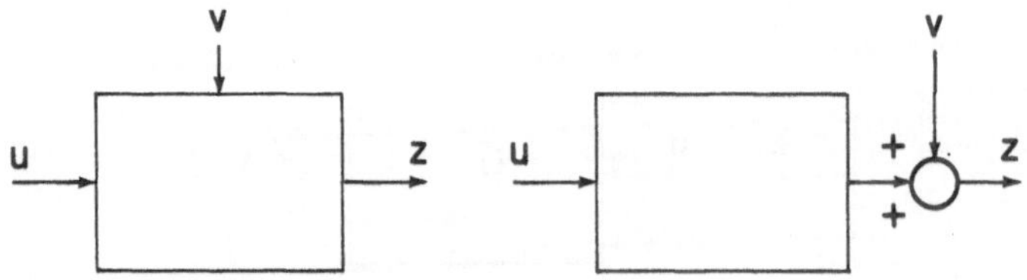

Fig. 5. Disturbances with specific point of entry (left) and equivalent disturbance collected at output (right).

(A) is sometimes referred to as structured uncertainty, and (B) as unstructured uncertainty. In case (B) there is no single or simple way of characterizing the plant uncertainty. One approach is to refine the analysis of the plant dynamics and add equations to the plant model that describe the marginal effects. In the extended plant model thus found the nominal model is obtained by setting one or several of the model parameters equal to zero. This is the starting point of the theory of singular perturbations within the framework of control theory, which has been a subject of some interest (Kokotovic, 1984).

Another approach to (B) is to assume a plant model of high complexity (usually this means a model of high order) with many unknown parameters that do not necessarily have a physical interpretation. The result is an uncertainty model of type (A). This line is sometimes followed in the study of adaptive systems.

Within the framework of single-input single-output linear time-invariant systems, plant uncertainty of the category (B) can sometimes usefully be represented by writing the perturbed frequency response function $H(i\omega)$ in the form $H(i\omega) = H_o(i\omega)P(i\omega)$, where H_o is the unperturbed (nominal) frequency response function, and P incorporates the effect of the perturbation. Nominally, $P(i\omega) = 1$. The possible size of the perturbations may be characterized in the form of bounds on the magnitude $|P(i\omega)|$ and the phase $\arg(P(i\omega))$ of P as a function of frequency ω.

3. Sensitivity and robustness

In the context of feedback control systems, originally the term sensitivity referred to the effect of both disturbances and perturbations on the control system output. During the late 1960's and early 1970's, the term sensitivity usually exclusively related to the effect of parameter variations (as in model (A)) on the control system performance (Cruz, 1973). Since the late 1970's the term robustness has come into use, with the meaning that the more robust a control system is, the less its performance is affected by perturbations, in particular unstructured perturbations. Therefore, the terms sensitivity and robustness to some extent are complementary.

For clarity we shall limit the use of the word sensitivity to the analysis of the effect of unwanted external signals (especially disturbances and measurement noise) on the control system output. The term robustness will only be used in the assessment of the effect of perturbations on the control system performance.

In the case of robustness it is in principle necessary to specify what kind of robustness is meant, since control system performance has several aspects. The most basic robustness requirement is stability robustness, which means that the control system remains stable under all possible perturbations. Performance robustness presumes stability robustness and indicates that certain performance specifications, such as on the rise-time of the step response and on the amount of disturbance attenuation, are not violated under all possible perturbations.

The remainder of this section will be devoted to an analysis of the sensitivity and robustness properties of the simplest type of control system, namely a single-input single-output feedback control system with two degrees of freedom, involving a compensator and prefilter, with the block diagram of Fig. 6. In the block diagram also the various external input signals are indicated, namely the disturbance v (collected at the plant output), the reference signal r, and the measurement noise w. The measurement noise arises as a result of feedback; although it is not always important its presence should nevertheless be accounted for. The plant transfer function is P, the compensator transfer function C, and the prefilter transfer function F. Taking a signal balance in terms of Laplace transforms at the output of the control system yields

$$Z = V + HG(FR - W - Z), \tag{5}$$

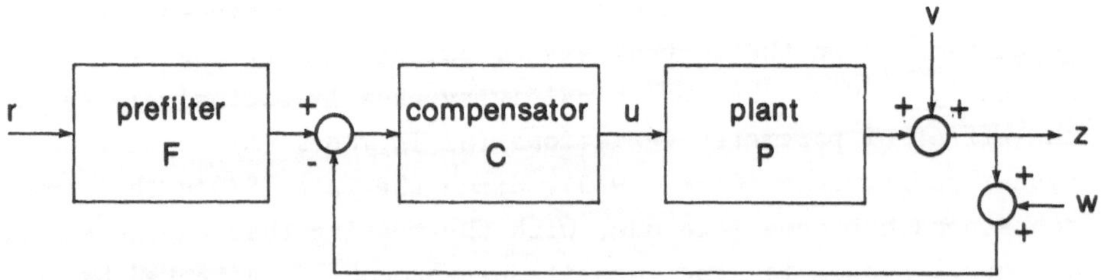

Fig. 6. Single-input single-output two-degree-of-freedom feedback system.

where Z, V, R and W denote the Laplace transforms of z, v, r and w, respectively. Solving for Z we obtain

$$Z = \frac{1}{I+HG}(V - HGW + HGPR). \tag{6}$$

Introducing the closed-loop system functions

$$S := \frac{1}{I+HG}, \quad T := \frac{HG}{I+HG}, \tag{7}$$

this can be rewritten as

$$Z = SV - TW + TPR. \tag{8}$$

The latter expression shows that the transfer function S determines how the disturbance v is transmitted to the control system output z; therefore S is often called the sensitivity function of the closed-loop system. We furthermore observe that the functions T and S are related as

$$S + T = 1. \tag{9}$$

Because of this relationship, T is sometimes called the complementary sensitivity function. If the prefilter is unity, T represents the transfer function from the reference signal r to the control system output z; for this reason it is sometimes also referred to as the closed-loop transfer function.

The expression (8) reveals very clearly that the sensitivity of the closed-loop system to disturbances is determined by the sensitivity function S, while the complementary sensitivity function T characterizes its sensitivity to measurement noise.

It will turn out that the control system robustness can also be characterized in terms of the system functions S and T. The analysis, however, is not as straightforward as sensitivity analysis. One reason is that the results depend on the nature of the information about the possible perturbations. We shall exemplify this by a stability robustness analysis for the case that the plant transfer function P and the compensator transfer function C are rational and given by

$$P = \frac{N}{D}, \qquad C = \frac{Y}{X}, \tag{10}$$

where N, D, Y and X are polynomials in the Laplace variable s. Then S and T are easily found to be given by

$$S = \frac{DX}{DX+NY}, \quad T = \frac{NY}{DX+NY}. \tag{11}$$

The common denominator DX+NY of these two transfer functions is the closed-loop characteristic polynomial. If all the roots of this polynomial are in the left-half complex plane, the closed-loop system is stable; otherwise it is unstable. It will be assumed that for the nominal plant $P_o = N_o/D_o$ the closed-loop system is stable; the question then is for what perturbations the closed-loop system remains stable.

We shall state three stability robustness tests under different assumptions concerning the perturbations. The derivations of the test rely on the principle of the argument from complex variable theory; the Nyquist stability theory originates from the same source.

(a) Suppose that the nominal plant denominator polynomial D_o and the perturbed denominator polynomial D have the same number of right-half plane roots. Then a sufficient condition for the perturbed closed-loop system to remain stable is that

$$|T_o(i\omega)| \cdot \left|\frac{P(i\omega) - P_o(i\omega)}{P_o(i\omega)}\right| < 1 \quad \text{for all } \omega. \tag{12}$$

Here $T_o := P_oC/(1+P_oC)$ is the nominal complementary sensitivity function. We observe that the quantity $(P(i\omega)-P_o(i\omega))/P_o(i\omega)$ is a measure for the relative plant perturbation at frequency ω. From this criterion one concludes that for good stability robustness $|T_o(i\omega)|$ should be small at those frequencies where the relative plant perturbation is large. This criterion is the single-input single-output version of the well-known stability robustness criterion of Doyle (1979).

(b) Suppose that the nominal plant numerator polynomial N_o and the perturbed plant numerator polynomial N have the same number of right-half plane roots. Then a sufficient condition for the perturbed closed-loop system to remain stable is that

$$|S_o(i\omega)| \cdot \left|\frac{P_o(i\omega)-P(i\omega)}{P(i\omega)}\right| < 1 \quad \text{for all } \omega. \tag{13}$$

Here $S_o := 1/(1+P_oC)$ is the nominal sensitivity function. In this criterion, $(P_o(i\omega)-P(i\omega))/P(i\omega)$ is also a measure for the relative plant perturbation at frequency ω, but slightly different from the one before. According to this criterion (under the given assumption), $|S_o(i\omega)|$ should be small at those frequencies where the relative plant perturbation is large.

We observe that the criteria (a) and (b) yield partly contradictory results when both N and N_o have the same number of right-half plane roots and also both D and D_o have the same number of right-half plane roots. When neither condition is satisfied no result is obtained at all. The next criterion (Kwakernaak, 1983) needs none of the two assumptions:

(c) A sufficient condition for the closed-loop system to remain stable is that

$$\left| S_o(i\omega)\,\frac{D(i\omega)-D_o(i\omega)}{D_o(i\omega)} + T_o(i\omega)\,\frac{N(i\omega)-N_o(i\omega)}{N_o(i\omega)} \right| < 1 \text{ for all } \omega. \tag{14}$$

This criterion involves the relative perturbation $(D-D_o)/D_o$ of the plant denominator as well as the relative perturbation $(N-N_o)/N_o$ of the plant numerator. Also, the criterion depends on both the nominal sensitivity S_o and the nominal complementary sensitivity function T_o.

For use in Section 5 we state the following immediate conclusions from the three robustness criteria.

(a') Suppose that all possible perturbations leave the number of right-half plane plant poles invariant and satisfy

$$\left| \frac{P(i\omega)-P_o(i\omega)}{P_o(i\omega)} \right| \le |W(i\omega)| \quad \text{for all } \omega, \tag{15}$$

with W a given frequency dependent function. Then the closed-loop system remains stable for any admissible perturbation if

$$|W(i\omega)T_o(i\omega)| < 1 \quad \text{for all } \omega. \tag{16}$$

(b') Suppose that all possible perturbations leave the number of right-half plane plant zeros invariant and satisfy

$$\left| \frac{P_o(i\omega)-P(i\omega)}{P(i\omega)} \right| \le |V(i\omega)| \quad \text{for all } \omega, \tag{17}$$

with V a given frequency dependent function. Then the closed-loop system remains stable under any admissible perturbation if

$$|V(i\omega)S_o(i\omega)| < 1 \quad \text{for all } \omega. \tag{18}$$

(c') Suppose that all possible perturbations satisfy

$$\left| \frac{(D(i\omega)-D_o(i\omega))/D_o(i\omega)}{V(i\omega)} \right|^2 + \left| \frac{(N(i\omega)-N_o(i\omega))/N_o(i\omega)}{W(i\omega)} \right|^2 < 1 \tag{19}$$

for all ω, where V and W are given frequency dependent functions. Then the closed-loop system remains stable for any admissible perturbation if

$$|V(i\omega)S_o(i\omega)|^2 + |W(i\omega)T_o(i\omega)|^2 < 1 \quad \text{for all } \omega. \tag{20}$$

For a discussion of the performance robustness of the control system of Fig. 6 we shall consider two transfer functions, namely the transfer function S (the sensitivity function) from the disturbances v to the control system output z, and the transfer function TF from the reference signal r to the output z. The hypothesis will be entertained that these two transfer functions determine the overall control system performance and hence that these two transfer functions should be affected as little as possible by plant perturbations.

To analyze the effect of a plant perturbation on S repectively TF, assume that the plant is perturbed from P_o to P. Then it is not difficult to find that the corresponding relative deviations of S from S_o and of TF from T_oF are given by

$$\frac{S-S_o}{S} = \frac{P_o-P}{P_o} T_o, \qquad \frac{TF-T_oF}{TF} = \frac{T-T_o}{T} = \frac{P-P_o}{P} S_o. \tag{21}$$

Again, these expressions involve the relative plant perturbations $(P_o-P)/P_o$ respectively $(P-P_o)/P$ and the nominal complementary sensitivity function T_o respectively the nominal sensitivity function S_o. For good performance robustness, these two functions should be kept small in appropriate frequency ranges.

For the time being, this completes our discussion of the sensitivity and robustness of control systems with the configuration of Fig. 6. There are other ways in which sensitivity and robustness can be assessed. Some of these will be encountered in Sections 4 and 5, which deal with the design of insensitive respectively robust control systems.

4. Design of insensitive control systems

Over the years, many techniques have been developed for the design of closed-loop control systems with useful sensitivity properties. The earliest techniques, now known as "classical" control systems design, center on single-input single-output linear time-invariant systems, described in the frequency domain. As we have seen in the previous section, low sensitivity to disturbances is obtained when the sensitivity function S is small over the frequency range of the disturbances. Therefore, classical design techniques (Horowitz, 1963) concentrate on making the loop gain L := PC large because this will result in a small sensitivity function

$$S = \frac{1}{1+PC} = \frac{1}{1+L} \quad . \tag{22}$$

A large loop gain, however, entails several potential risks:

(a) A large loop gain may not be compatible with stability. Indiscriminately increasing the loop gain will make almost any closed-loop system unstable. With some plants (in particular those with right-half plane zeros) the gain can only be made large in certain frequency regions at the expense of a small gain and accordingly large sensitivity in other regions.

(b) Large loop gains may lead to unacceptable large plant inputs so that the plant capacity is exceeded.

(c) Large loop gains may result in unacceptably large sensitivity to measurement noise.

Classical control system design deals with these problems in a more or less ad hoc manner. Graphical techniques prevail. The manipulated object usually is the loop gain $L(i\omega)$. The mathematics that is needed largely amounts to complex analysis.

Control system design theory assumed a different appearance with the advent of optimal control. As long as optimal control remained deterministic, it obscured the need for feedback. When stochastic models are used for the definition of optimal control problems, feedback can no longer be evaded, because the stochastic nature of disturbances necessitates the continuous observation of the plant state or its estimation from current output observations.

Within the context of optimal control system design, until recently only two techniques have emerged that can deal with disturbance and measurement noise attenuation in a way that has some promise for practical engineering applications. The first of these is Wiener optimization. This approach was developed in the 1950's based on Wiener's optimal filtering theory (Newton, Gould and Kaiser, 1957). Although this work had much influence on the thinking of control theorists it never really led to large scale practical application. One reason was perhaps that Wiener optimization, like other sophisticated design methods, needs well-developed CAD tools that at the time were still far from view.

Another reason that Wiener optimization was dropped, at least temporarily, was that it was superseded by what is often known as

"LQG" design (Athans and Falb, 1966; Anderson and Moore, 1971; Kwakernaak and Sivan, 1972). This approach is based on an elegant mathematical theory and uses the state space approach, which for many applications has much physical appeal. LQG owes much to Kalman's work. The starting point for this theory is a linear state space model of the form

$$\begin{aligned} \dot{x} &= Ax + Bu + Gv, \\ z &= Dx, \\ y &= Cx + w. \end{aligned} \tag{23}$$

Here x is the state of the plant, u the input, z the controlled output, y the observed output, v the disturbance, and w the measurement noise. All these signals can be vector-valued; the fact that such signals can easily be handled was at the time a considerable advantage over the Wiener theory. For both the disturbance v and the measurement noise w Gaussian white noise models are assumed. A, B, G, C and D are constant coefficient matrices.

The design of a control system for the plant (23) is converted to an optimization problem where one wishes to minimize a criterion of the form

$$E\{\int_{t_o}^{t_1}[z(t)^TQz(t) + u(t)^TRu(t)]dt\}. \tag{24}$$

In this expression, t_o is the starting time, t_1 the final time, Q and R are constant symmetric weighting matrices, and E denotes the mathematical expectation. The first term in the criterion represents the control quality, the second the necessary plant input power. The problem formulation is that of a regulation problem, with no external reference input. It explicitly includes disturbances and measurement noise as well as a plant power constraint. The problem is denoted LQG because the plant model is Linear, the criterion Quadratic, and the noises are Gaussian. The problem formulation is wide and adaptible enough to fit a large number of practical situations.

The solution is essentially of a feedback nature: the optimal control is given by

$$u = -F\hat{x}. \tag{25}$$

Here F is a gain matrix which in the limit $t_1 \to \infty$ is independent of t, and $\hat{x}$ is the optimal estimate of the state x of the plant as produced by an on-line Kalman filter (Fig. 7). The equation of the Kalman filter is

$$\dot{\hat{x}} = A\hat{x} + Bu + K(y-C\hat{x}), \tag{26}$$

where the filter gain matrix K again in the limit $t_1 \to \infty$ does not depend on t. The gains F and K follow by solving two matrix Riccati equations that will not be given here.

LQG has made some headway into sophisticated applications, in particular aerospace. Much effort has been spent on refining and extending the theory. Many useful properties of LQG-optimal control systems have been discovered, some of which will be mentioned in the next section.

5. Design of robust control systems

The classical approach to the design of robust control systems ties in with the graphically oriented approach mentioned in the preceding section. In a closed-loop system where neither the plant nor the compensator have right-half plane poles the closed-loop system -- by the Nyquist criterion -- is stable if and only if the Nyquist plot of the loop gain L = PC does not encircle the point -1 (Fig. 8). Control systems where the Nyquist plot of the loop gain does not encircle -1 but approaches this point closely exhibit an undesirable near-oscillatory behavior. Moreover, if the Nyquist plot closely passes by the point -1, small perturbations may easily make the closed-loop system unstable. The classical gain and phase margins constitute a

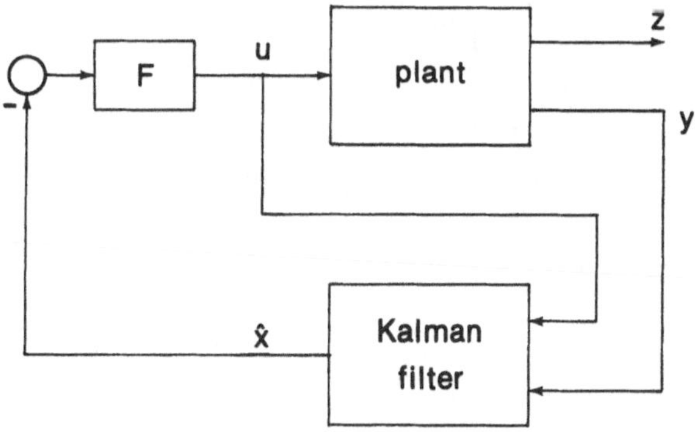

Fig. 7. Structure of an LQG-optimal control system.

semi-quantitative measure of the extent to which the Nyquist plot of the loop gain avoids the point -1. The gain margin is the factor k by which the loop gain may be multiplied such that its Nyquist plot goes through the point -1. The phase margin is the amount of phase α that can be subtracted from the loop gain such that its Nyquist plot goes through -1 (see Fig. 8).

Classical design techniques concentrate on achieving a maximal loop gain over a specified frequency range consistent with specified gain and phase margins. Unfortunately these techniques often are inadequate for nonminimum-phase plants, i.e. plants that have right-half plane poles or zeros or both. Moreover, multi-input multi-output plants can not really be handled.

Horowitz's work descends directly from the classical line. He has developed semi-analytical semi-graphical techniques (Horowitz, 1979) for the design of robust single- and multi-loop feedback systems that rely on specifying and achieving frequency dependent bounds on important closed-loop system transfer functions.

When we stop a moment to think about what optimal control can offer for the design of robust and insensitive control systems, some contemplation reveals that in principle stochastic optimal control could solve all problems (Bellman, 1961). The fact that it does not do so for problems of more than academic interest can fully be blamed on the inadequacy of current numerical methods and computing power. To set up the problem in a stochastic optimal control framework one has to be a believer in Bayesian statistics. Suppose that in state space form the

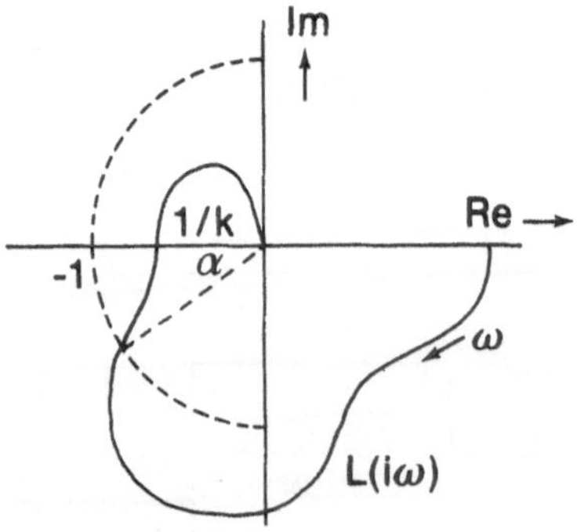

Fig. 8. Nyquist plot of the loop gain of a control system with gain margin k and phase margin α.

plant can be represented in the following way (we choose a discrete-time description but an analogous continuous-time model can be formulated as well)

$$x_{t+1} = f(x_t, p, u_t, w_t, t),$$
$$y_t = g(x_t, p, u_t, w_t, t). \tag{27}$$

Here x_t is the state of the plant, u_t the plant input, y_t the observed plant output and p a set of unknown plant parameters. Furthermore, w_t is a (vector-valued) random process that represents the effect both of disturbances and measurement noise. In postulating this model it is assumed that all plant uncertainty is parametrized by p, which may be of high dimension indeed. The next step is to incorporate p into an enlarged state vector $x_t' := (x_t, p)$. To make the stochastic model complete, an initial probability distribution is required for the enlarged state x_t'. For the original state x_t this probability distribution is assumed to be available, for the added component p a (subjective) distribution is constructed on the basis of all available a priori information.

The final step that is needed to obtain a conventional stochastic control problem is to establish an optimization criterion of the form

$$E\{\sum_{t=t_o}^{t_1} h(x_t', u_t, w_t, t)\}, \tag{28}$$

with h a suitable function, t_o the initial time and t_1 the final time, and E the mathematical expectation. This criterion should include both the performance quality and constraints on the plant input power and on any other variables.

Stochastic optimal control has reached a level of high sophistication and many results are available on the existence of solutions. Unfortunately, the actual construction of solutions, except in the simplest cases, involves the recursive resolution of a Bellman functional equation of infinite dimension which at this time is beyond our resources.

Some thought has been given to properties that the solution of such problems may or may not possess. In the early 1960's, Feldbaum (1965) published some work which contemplates this. He pointed out that part of the information processing that takes place within the optimal feedback controller should be directed towards the identification of the unknown plant parameters. He called this dual control. He conjectured that in certain situations part of the plant input power might solely be devoted to improving the identification so that in the long run better control is achieved, although on a short-term basis the effects might be adverse. This is called exploratory control. Other notions that were coined in this context were for instance neutral control systems (systems where the accuracy of the identification is not influenced by the plant input behavior and hence exploratory control is not useful), certainty equivalence (situations where the form of the control law is not influenced by the introduction of uncertainty), the separation principle (this applies in situations where the estimation of the state and the parameters can be completely separated from the determination of the optimal control), and caution (this means that in case of uncertainty potentially risky control actions are avoided).

In Fig. 9 a block diagram is given of a "dual" feedback control system which explicitly includes an identification loop. In fact this is nothing else but the familiar structure of an adaptive control system, where the identification loop tracks any plant parameter changes and the compensator is accordingly adjusted. Adaptation is a appealing way of obtaining robust control system behavior, and is the subject of another lecture during this seminar (Åström, 1985).

We next turn to a discussion of LQG. Although very well equipped to deal with reducing sensitivity to disturbances and measurement noise, LQG regretfully has failed to provide fully adequate tools for the design of robust control systems. On first inspection, though, LQG-

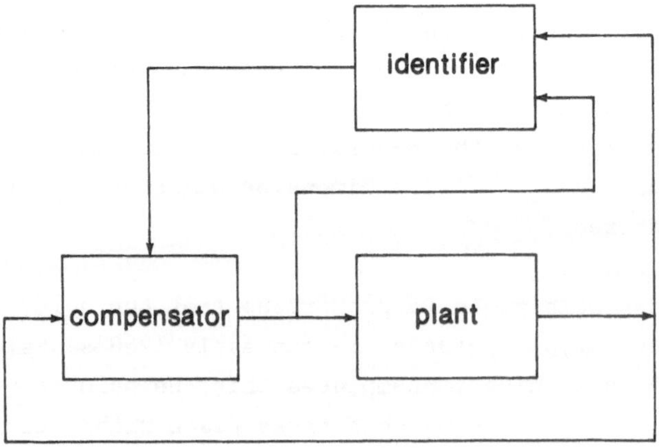

Fig. 9. "Dual" or adaptive control system.

optimal control systems appear to have properties that promise well for robustness. These stem from the fact that as we have seen in Section 3 there are close connections between insensitivity and robustness. We state some of these results for the case of single-input plants; they also apply, with appropriate modification, to the multi-input case (Lektomaki, Sandell and Athans, 1981).

Consider LQG-optimal control with full state feedback. This means that the complete state can be accurately measured at all times. In this case the observation equation in (23) takes the form $y = x$, and the optimal control law (25) becomes

$$u = -Fx, \tag{29}$$

rather than $u = -F\hat{x}$. Since the plant transfer function is $P = (sI-A)^{-1}B$ and the compensator transfer function is $C = F$, we can define a loop gain $L := CP = F(sI-A)^{-1}B$, which in the case of a single-input system is a scalar function. It then can immediately be proved (Kalman, 1964) from the Riccati equation from which F is obtained that

$$|1+L(i\omega)| > 1 \quad \text{for all } \omega. \tag{30}$$

This means that the Nyquist plot of L lies completely outside the circle with center -1 and radius 1, thus providing generous gain and phase margins (see Fig. 10). This result also implies that if we define the sensitivity-like function

$$S := \frac{1}{I+L} \quad , \tag{31}$$

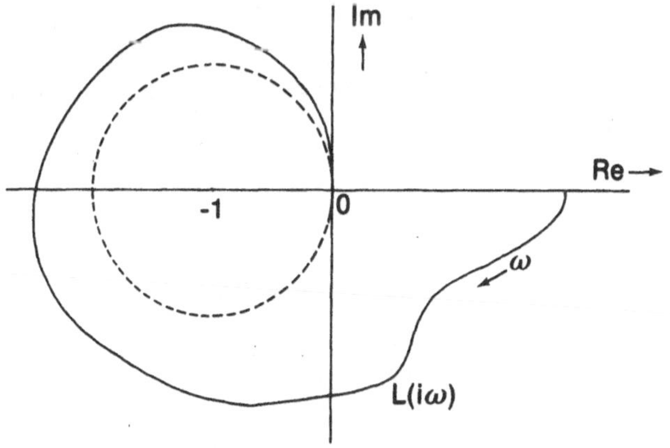

Fig. 10. Sample Nyquist plot of the loop gain of an LQG-optimal control system

we have

$$|S(i\omega)| < 1 \quad \text{for all } \omega. \tag{32}$$

One problem with these results is that they apply to a single-input multi-output plant where for the definition of the loop gain L the loop is cut at the plant input rather than the plant output, as needed for a useful interpretation.

With some effort, however, the results just mentioned can be interpreted as useful properties for closed-loop insensitivity and robustness, but worse is that first of all they turn out not to hold for sampled-data LQG-optimal systems, and secondly they also fail to hold in the case of output feedback. Since state feedback is only exceptionally feasible and output feedback is the norm, this makes the results that we described lose much of their value.

Some consolation derives from the fact that the properties (30) and (32) in the case of output feedback are asymptotically recovered for plants that have no right-half plane zeros if either the weighting matrix R in the criterion (24) goes to zero (Kwakernaak, 1969) or the measurement noise intensity goes to zero (Doyle and Stein, 1979). These results have been used by Doyle and Stein in their so-called loop transfer recovery (LTR) procedure for the design of robust multivariable linear feedback systems (Doyle and Stein, 1981).

The difficulties in applying LQG to robust control system design stem from the fact that the LQG problem formulation does not directly include robustness requirements. Some efforts have been made to extend LQG in this respect. This may be done by adding extra terms to the criterion representing the effect of perturbations, usually in the form of induced disturbances (Bongiorno 1969; Youla and Bongiorno, 1985).

Another line has been taken by Ackermann (1983). In this work, the observation is exploited that the closed-loop response and other properties of a linear feedback system usually stay more or less invariant as long as the dominant closed-loop poles stay within a prescribed area in the complex plane. Combined analytical-graphical techniques are used to obtain state feedback laws with the property that the closed-loop poles remain in a given area for all admissible parameter variations. Subsequently the control law is modified so that the same result is obtained for output feedback. This approach is related to the so-called simultaneous stabilization problem, where the task is set to find a single compensator that stabilizes any plant of a given family of plants. A more extensive discussion is given in another contribution to this seminar (Ackermann, 1985).

In recent years the inadequacy of LQG to cope with robustness problems has been blamed on the fact that LQG employs an integral criterion (Zames and Francis, 1983). In the state space presentation of LQG of Section 4 this integral criterion is formulated in the time-domain. In the case of time-invariant plants and control periods of infinite duration, Parseval's theorem can be used to convert the time-domain criterion into a frequency-domain criterion, which in the case of a single-input single-output plant is of the form

$$\int_{-\infty}^{\infty} [|S(i\omega)|^2 + |W(i\omega)T(i\omega)|^2]\Phi(\omega)d\omega \quad . \tag{33}$$

Here Φ is the power spectral density of the disturbances collected at the plant output and W is an appropriate frequency dependent weighting function. Minimization of this criterion does not force $|S(i\omega)|$ and $|T(i\omega)|$ to be <u>uniformly</u> small at all frequencies; local peaks of $|S(i\omega)|$ and $|T(i\omega)|$, provided their area is sufficiently small, can not be excluded. Now as we have seen in Section 3, the stability robustness criteria that were formulated require $|S(i\omega)|$ and $|T(i\omega)|$, possibly appropriately weighted, to be uniformly small for all frequencies. In fact it is easy to construct examples where for a given perturbation an arbitrarily narrow but sufficiently high peak in $|S(i\omega)|$ or $|T(i\omega)|$ destroys the stability of the closed-loop system.

This argument has been used to defend a new optimization technique which ensures $|S(i\omega)|$ and/or $|T(i\omega)|$ to be uniformly small or at least bounded. Although there seems to be no example of an LQG design with specifically unfavorable robustness properties, the new optimization theory is appealing because it allows the direct inclusion of robustness requirements, which is not easily done in LQG.

To further motivate the minimax optimization theory that we are about to introduce let us recall the robustness criteria of Section 3. The three criteria required the expressions $|V(i\omega)S(i\omega)|$, respectively $|W(i\omega)T(i\omega)|$, respectively $|V(i\omega)S(i\omega)|^2 + |W(i\omega)T(i\omega)|^2$ to be uniformly less than 1. Here V and W are given weighting functions, determined by the relative size of the weighting functions. Conversely, if we could find a compensator such that $|V(i\omega)S(i\omega)|$

respectively $|W(i\omega)T(i\omega)|$ respectively $|V(i\omega)S(i\omega)|^2 + |W(i\omega)T(i\omega)|^2$ are uniformly less than ε, with $0 < \varepsilon < 1$, the functions V and W could be made larger, thus ensuring robust stability for a larger class of perturbations. From this consideration, choosing the compensator such that any of the three following criteria is minimized, depending on the a priori assumptions,

$$\begin{aligned} &\sup_\omega \ |V(i\omega)S(i\omega)|, \qquad \sup_\omega |W(i\omega)T(i\omega)|, \\ &\sup_\omega \ [|V(i\omega)S(i\omega)|^2 + |W(i\omega)T(i\omega)|^2] \end{aligned} \tag{34}$$

could be termed robustness optimization.

However, robustness is not the only aspect of control system evaluation. We shall for brevity only account for two further aspects, namely disturbance sensitivity and plant input power limitations. Disturbance sensitivity can be restricted if an expression of the form $|V'(i\omega)S(i\omega)|$, with V' a suitable weighting function (large at those frequencies where there may be large disturbances) is uniformly small over all frequencies. As for plant input power, it is easily verified that in the block diagram of Fig. 6 the transfer function from the disturbance v to the plant input u is given by T/P. Thus, plant input power can be restricted if an expression of the form $|W'(i\omega)T(i\omega)|$ is kept bounded, where W' is another weighting function that accounts both for the factor 1/P in the transfer function and the spectral distribution of the disturbances.

Combination of the preceding considerations leads to the minimization of a criterion of the form

$$\sup_\omega \ [|V(i\omega)S(i\omega)|^2 + |W(i\omega)T(i\omega)|^2] \tag{35}$$

where the weighting functions V and W are so chosen that they reflect both the robustness and performance requirements (Kwakernaak, 1983).

This minimax optimization problem is currently receiving considerable attention (see e.g. Zames and Francis, 1983; Francis, Helton and Zames, 1984; Kwakernaak, 1983, 1984). It often is referred to as H^∞-optimization, after the mathematical space in which the problem can be placed. Its solution, especially the multivariable version, requires a considerable mathematical apparatus with techniques that for the most part are new to the systems and control community. Although quite surprisingly the analytical development can be taken very far, it looks as if numerically the problem is also rather demanding. However, the subject is in full development and the end is not in sight. A promising feature is that many of the classical control system design objectives can be directly incorporated into the problem formulation. At this time it is a little early to tell how useful the methodology will be for practical control system design.

6. Conclusion

Uncertainty is a central issue in feedback control system design. In fact, uncertainty is the main reason for feedback. Although in the recent history of control theory this has sometimes been lost from sight, it has always returned to the forefront.

Currently there are two approaches to the problem of designing feedback systems that can cope with large degrees of uncertainty. The first is the theory of adaptive control systems. It is left to the experts on this subject to evaluate its practical potential (Åström, 1985).

The other approach is robust control system design. This term covers a variety of results and theories. The most pretentious of these no doubt is H^∞- or minimax frequency domain optimization. From a practical engineering standpoint, the classical techniques still are the most useful. The younger generation control engineers, most of whom are familiar with LQG now, may find some use for some ideas about robustness that have been developed in this context. How practical H^∞-optimization is still remains to be seen.

All techniques that are being developed or considered for robust control system design involve a fair to considerable amount of manipulation and numerical effort, especially for multivariable systems. For this reason, any reasonably sophisticated design methodology can only be expected to be practically successful if it is integrated within a very comfortable and powerful CAD package. Moreover, the design philosophy should be clear and straightforward and the CAD package should reflect this lucidity. No such package exists at this time and, moreover, control engineers have not been educated to use them. Recently there has been an emerging interest in the development of CACSD (computer aided control system design) packages. This is as much a control engineering as a computer science task and it may take us well into the fifth generation computer era until it is completed.

References

J. Ackermann (1983). Abtastregelung, Bd. II, Entwurf robuster Systeme. Springer-Verlag, Berlin.

J. Ackermann (1985), "Multi-model approaches to robust control system design". This volume.

B.D.O. Anderson and J.B. Moore (1971). Linear Optimal Control. Prentice-Hall, Englewood Cliffs.

K.J. Åström (1985), "Adaptive control -- a way to deal with uncertainty". This volume.

M. Athans and P.L. Falb (1966). Optimal Control. McGraw-Hill, New York.

R. Bellmann (1961). Adaptive Control Processes, A Guided Tour. Princeton University Press, Princeton.

J.J. Bongiorno (1969), "Minimum sensitivity design of linear multivariable feedback control systems by matrix spectral factorization". IEEE Trans. Aut. Control 14, pp. 665-673.

J.B. Cruz, Jr., Ed. (1973). System Sensitivity Analysis. Benchmark Papers in Electrical Engineering and Computer Science. Dowden, Hutchinson and Ross, Stroudsburg.

J.C. Doyle and G. Stein (1979), "Robustness with observers". IEEE Trans. Aut. Control 24, pp. 607-611.

J.C. Doyle (1979), "Robustness of multiloop linear feedback systems". Proc. 17th IEEE Conference on Decision and Control, pp. 12-18.

J.C. Doyle and G. Stein (1981), "Multivariable feedback design: Concepts for a classical/modern synthesis". IEEE Trans. Aut. Control 26, pp. 4-16.

A.A. Fel'dbaum (1965). Optimal Control Systems. Academic Press, New York.

B.A. Francis, J.W. Helton and G. Zames (1984), "H^∞-optimal feedback controllers for linear multivariable systems". IEEE Trans. Aut. Control 29, pp. 888-900.

I.M. Horowitz (1963). Synthesis of Feedback Systems. Academic Press, New York, 1963.

I.M. Horowitz (1979), "Quantitative synthesis of uncertain multiple input-output feedback systems". Int. J. of Control 30, pp. 589-600.

R.E. Kalman (1964), "When is a linear system optimal?". Trans. ASME, J. Basic Eng. 86, pp. 51-60.

P. Kokotovic (1984), "Applications of singular perturbation techniques to control problems", SIAM Review 26, pp. 1-90.

H. Kwakernaak (1969), "Optimal low-sensitivity linear feedback systems". Automatica 5, pp. 279-285.

H. Kwakernaak and R. Sivan (1972). Linear Optimal Control Systems. Wiley-Interscience, New York.

H. Kwakernaak (1983), "Robustness optimization of linear feedback systems". Proc. 22nd IEEE Conference on Decision and Control, San Antonio.

H. Kwakernaak (1984), "Minimax frequency domain optimization of multivariable linear feedback systems". Proc. 9th IFAC World Congress, Budapest, Pergamon Press, Oxford.

N.A. Lektomaki, N.R. Sandell and M. Athans (1981), "Robustness results in LQG-based multivariable control systems design". IEEE Trans. Aut. Control 26, pp. 75-93.

G.C. Newton, L.A. Gould and J.F. Kaiser (1957). Analytical Design of Linear Feedback Controls. Wiley, New York.

D.C. Youla and J.J. Bongiorno (1985), "A feedback theory of two-degree-of-freedom optimal Wiener-Hopf design". To appear, IEEE Trans. Aut. Control, 30.

G. Zames and B.A. Francis (1983), "Feedback, minimax sensitivity, and optimal robustness", IEEE Trans. Aut. Control 28, pp. 585-601.

MULTI-MODEL APPROACHES TO ROBUST CONTROL SYSTEM DESIGN

J. Ackermann
DFVLR-Institut für Dynamik der Flugsysteme
8031 Oberpfaffenhofen, FRG

Abstract

The problem of control system design is stated with explicit uncertainty bounds for physical parameters in the plant model and performance bounds as design objectives. A finite number of typical plant parameter values is used to define a multi-model problem. Two design methods for fixed-gain controllers for this problem are reviewed: i) the simultaneous assignment of the poles to a given region for all members of the plant family by parameter space methods, ii) the interactive Pareto-optimization of a vectorial performance index. The controller for the representative family of plant models may then be tested for continuous intervals of the parameter uncertainties. Only few results are available on stability of interval polynomials and matrices. Also the problem of a systematic choice of a feasible controller structure has not yet been solved. The interactive multi-model approaches will particularly profit from the progress in computer graphics.

1. Introduction, Examples

Most plant models used for controller design are uncertain. Even if an exact model is available, it may be so complicated that it must be approximated by a simpler, but uncertain, design model. For example, nonlinear models may be linearized for small deviations from an operating condition. Then the linear model depends on this uncertain operating condition. Also physical parameters of the plant and its environment may be uncertain. Suppose the linearized plant can be described by a state space model

$$\begin{aligned} \dot{\underline{x}} &= \underline{A}(\underline{\Theta})\underline{x} + \underline{B}(\underline{\Theta})\underline{u} \\ \underline{y} &= \underline{C}\underline{x} \end{aligned} \qquad (1)$$

where $\underline{\Theta}$ is the vector of uncertain plant parameters. Assume the state variables in $\underline{x}$ are chosen such that the output matrix $\underline{C}$ does not depend on $\underline{\Theta}$.

It is desired to find a feedback control law, i.e. a functional

$$\underline{u}(t) = F(\underline{y}(\tau), \underline{\Theta}) \qquad \tau \le t \tag{2}$$

such that the closed loop

i) is nicely stable. In the case when $\underline{\Theta}$ is constant this means the closed-loop system has rapidly decaying and well-damped modes,

ii) has required disturbance compensation, filtering and tracking properties,

iii) requires only admissible actuator signal magnitudes, i.e. $|u| \le u_{max}$,

iv) has the properties i), ii) and iii) for all constant values of the plant parameters $\underline{\Theta}$ in a given admissible set Ω, i.e. for $\underline{\Theta} \in \Omega$.

The simplest control law (2) is a linear controller. A linear dynamic controller structure may be assumed in form of a state-space model or transfer function with input $\underline{y}$ and output $\underline{u}$. The free design parameters in this structure are combined to a controller-parameter vector $\underline{k}'$.

We call this a fixed-gain controller if $\underline{k}$ is constant. In a gain scheduling controller, some components of $\underline{\Theta}$ or related external variables are measured and used for adjustment of $\underline{k} = \underline{k}(\underline{\Theta})$. In an adaptive controller, $\underline{k}$ is a functional $\underline{k}(t) = F(\underline{u}(\tau), \underline{y}(\tau))$, $\tau \le t$. Adaptive control systems are reviewed by Åström [30] and Tsypkin [31]. In this paper the design of fixed-gain, robust controllers will be reviewed.

Example 1: Track-guided bus, [1], [2], [3]

A bus is guided by the field generated by a wire in the street. Fig. 1 shows the case where the nominal track is a straight guideline.

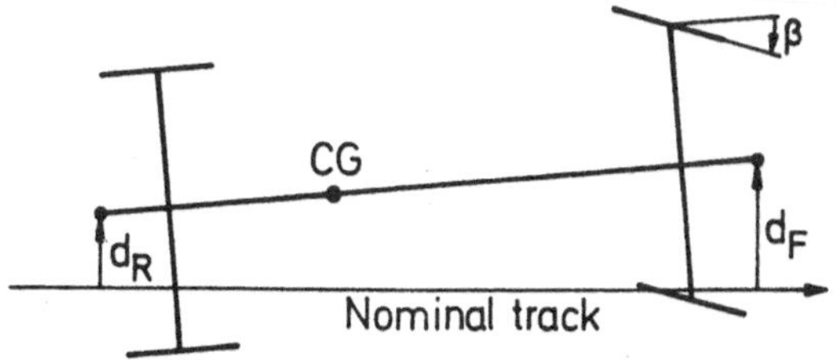

Fig. 1
Track-guided bus
δ_R, δ_F = rear and front displacement of the bus from the guideline,
β = steering angle.

A feasible state vector is

$$\underline{x} = [d_F \quad \dot{d}_F \quad d_R \quad \dot{d}_R \quad \beta]' \tag{3}$$

A good controller will keep all state variables small, thus the controller design may be performed with the linearized model

$$\dot{\underline{x}} = \begin{bmatrix} 0 & 1 & 0 & 0 & 0 \\ a_{21} & a_{22} & a_{23} & a_{24} & a_{25} \\ 0 & 0 & 0 & 1 & 0 \\ a_{41} & a_{42} & a_{43} & a_{44} & a_{45} \\ 0 & 0 & 0 & 0 & -b_5 \end{bmatrix} \underline{x} + \begin{bmatrix} 0 \\ 0 \\ 0 \\ 0 \\ b_5 \end{bmatrix} u \tag{4}$$

$b_5 = 4.7s^{-1}$ is the time constant of the power steering. The coefficients a_{ij} are of the form

$$a_{ij} = \begin{cases} \alpha_{ij} \frac{\mu}{m} + \beta_{ij} \frac{\mu}{J} & \text{for } j = 1, 3, 5 \\ \alpha_{ij} \frac{\mu}{mv} + \beta_{ij} \frac{\mu}{Jv} & \text{for } j = 2, 4 \end{cases} \tag{5}$$

The coefficients α_{ij} and β_{ij} depend only on the bus geometry and are known. The variable plant parameters are

m = mass
J = moment of inertia
v = velocity
μ = road adhesion coefficient.

For uniform passenger distribution, J is determined by m and the plant parameter vector is

$$\underline{\Theta} = \begin{bmatrix} m \\ v \\ \mu \end{bmatrix} \tag{6}$$

The parameter ranges are:

$$\begin{aligned} m \text{ (empty bus)} &\leq m \leq m \text{ (full bus)} \\ v_{min} &\leq v \leq v_{max} \\ \mu \text{ (wet road)} &\leq \mu \leq \mu \text{ (dry road)} = 1 \end{aligned} \tag{7}$$

Fig. 2 illustrates the parameter ranges of m and v.

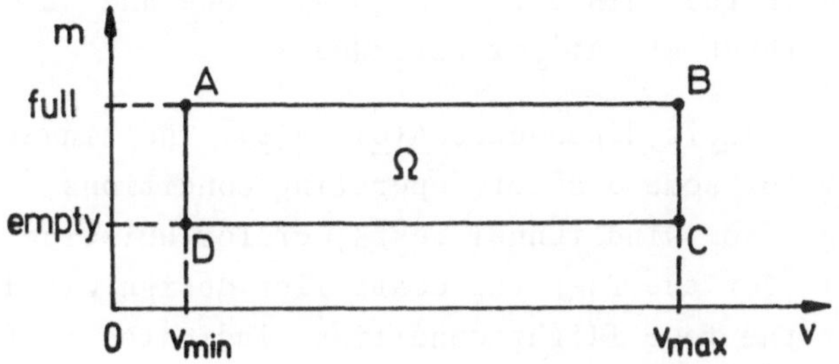

Fig. 2 Plant-parameter plane with coordinates v and m. Admissible region Ω

The measured variables are d_F, d_R and β, thus

$$\underline{y} = \underline{C}\underline{x} = \begin{bmatrix} 1 & 0 & 0 & 0 & 0 \\ 0 & 0 & 1 & 0 & 0 \\ 0 & 0 & 0 & 0 & 1 \end{bmatrix} \underline{x} \tag{8}$$

Track-guided buses are being developed in Germany by MAN and Daimler-Benz.

Example 2: Aircraft stabilization, [4], [5], [6]

The mathematical model of an aircraft is nonlinear. It may be linearized for small deviations from stationary flight with constant altitude h and velocity v. The linearized model thus depends on the plant parameter vector

$$\underline{\Theta} = \begin{bmatrix} v \\ h \end{bmatrix} \tag{9}$$

Fig. 3 shows the admissible region Ω of this parameter vector for an experimental aircraft F4-E with additional canards

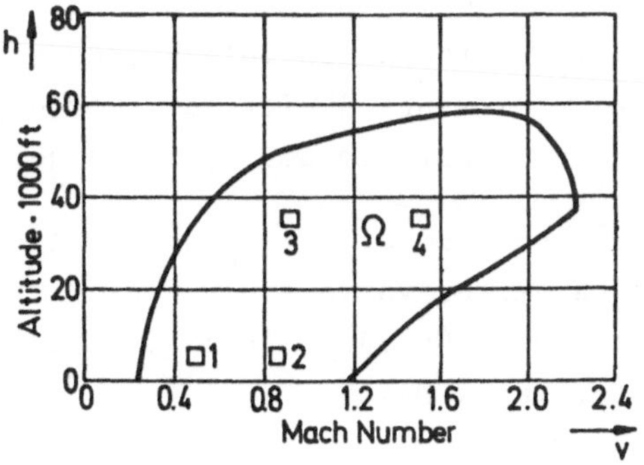

Fig. 3
Envelope of the possible flight conditions and four representative flight conditions

In this example the bounds in v and h are not independent. (E.g. the aircraft cannot fly slow in the thin air at high altitude and it cannot fly very fast in the thick air at low altitude.)

Also we do not know the explicit dependence $\underline{A}(\underline{\Theta})$, $\underline{B}(\underline{\Theta})$. The linearized equations are known only for some discrete operating conditions $\underline{\Theta} = \underline{\Theta}_j$, j = 1,2...N (e.g. from wind tunnel tests, or for existing aircraft from flight tests). For the F4-E the controller design was done at McDonnell Douglas for the four flight conditions indicated in fig. 3, [4], [5]. For a similar aircraft, the Swedish JAS 39, the controller design is done at Saab for ten flight conditions, which include also different locations of the center of gravity [6], [32].

2. Problem Formulation

A standard problem formulation uses the idea of a nominal plant $\underline{A}(\underline{\Theta}_o)$, $\underline{B}(\underline{\Theta}_o)$ and the sensitivity of the control-loop performance with respect to small perturbations $\Delta\underline{\Theta}$ from $\underline{\Theta}_o$. The design goal is that the closed-loop dynamics for $\underline{\Theta} = \underline{\Theta}_o + \Delta\underline{\Theta}$ are not much different from those for the nominal plant parameters $\underline{\Theta}_o$. Usually this is combined with the optimization of a performance index P for the nominal $\underline{\Theta}_o$.

Fig. 4 shows a design 1 with a good tradeoff between the requirements for minimal $P(\Theta_o)$ and minimal $\left.\frac{dP(\Theta)}{d\Theta}\right|_{\Theta = \Theta_o}$.

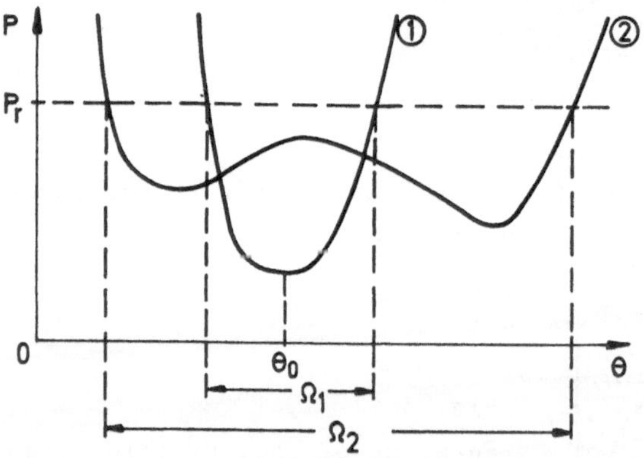

Fig. 4
In design 1 the performance index P at Θ_o and its local sensitivity with respect to Θ is optimal. In design 2 the admissible parameter range Ω_2 for satisfactory performance P_r is larger.

This local low sensitivity in the neighborhood of $\Theta = \Theta_o$ does not guarantee, however, that the admissible parameter variation Ω_1 for a required satisfactory performance P_r is good. Design 2 in fig. 4 results in a larger admissible plant parameter range Ω_2.

If no specific uncertainty structure is known for the model, then a standard structure is assumed for perturbations of the nominal plant $\underline{A}_o = \underline{A}(\underline{\Theta}_o)$, $\underline{B}_o = \underline{B}(\underline{\Theta}_o)$, e.g.

$$\begin{aligned} \dot{\underline{x}} &= (\underline{A}_o+\Delta\underline{A})\underline{x} + (\underline{B}_o+\Delta\underline{B})u \\ \underline{y} &= \underline{Cx} \end{aligned} \tag{10}$$

or perturbations of the corresponding transfer function

$$\underline{G}_o = \underline{C}[s\underline{I}-\underline{A}_o]^{-1}\underline{B}_o \tag{11}$$

There are three standard forms of such perturbations $\Delta\underline{G}$ in the frequency domain

i) additive perturbation $\underline{G}_o + \Delta\underline{G}$ (12)

ii) multiplicative perturbation $\underline{G}_o(\underline{I}+\Delta\underline{G})$ (13)

iii) feedback perturbation $\underline{G}_o(\underline{I}+\underline{G}_o\Delta\underline{G})^{-1}$ (14)

Typically the influence on stability of a class of small $\Delta\underline{G}$'s in the frequency domain is studied. Such methods are reviewed by Kwakernaak [29].

In this chapter we consider the problem of structured perturbations as illustrated by the examples "bus" and "aircraft". It is not recommended that such a problem be imbedded into a more general class of perturbations as in eqs. (10), (12), (13) or (14). In eq. (4) for example a perturbation $\Delta\underline{\Theta}$ will not change the "zero" and "one" elements of $\underline{A}$, $\underline{B}$ and the a_{ij}'s cannot vary independently. This has for example the effect that the actuator eigenvalue at s = -4.7 and the double eigenvalue at s = 0 remain unchanged under large perturbations of m, v and μ. Methods admitting general perturbations will naturally lead to more conservative bounds on the admissible perturbations $\Delta\underline{A}$, $\Delta\underline{B}$ or $\Delta\underline{G}$ and therefore on $\Delta\underline{\Theta}$.

We want to make use of the knowledge about the structure of the perturbation. In contrast to sensitivity methods we seek satisfactory performance for a large parameter range Ω, see Ω_2 in fig. 4. In fact, the introductory examples show that Ω may be given, thus Ω should be considered systematically in the design and not only analyzed a posteriori.

A typical basic problem is that of stability. The coefficients of the closed-loop characteristic polynomial are functions of both the plant parameters $\underline{\Theta}$ and the controller parameters $\underline{k}$, i.e.

$$P(s,\underline{\Theta},\underline{k}) = p_0(\underline{\Theta},\underline{k})+p_1(\underline{\Theta},\underline{k})s+\ldots+p_{n-1}(\underline{\Theta},\underline{k})s^{n-1}+s^n \tag{15}$$

A typical robustness problem is then: Find a $\underline{k}$ such that the roots of $P(s,\underline{\Theta},\underline{k})$ have negative real parts for all $\underline{\Theta}\in\Omega$. Or more generally: Find the set of all such $\underline{k}$ if any exist. This problem can be visualized in the combined space of $\underline{\Theta}$ and $\underline{k}$. In this space a stability region can be determined, which contains all $(\underline{\Theta},\underline{k})$ such that (15) is a Hurwitz polynomial. There are two possibilities to break down the problem into two lower dimensional ones:

i) For fixed $\underline{k}$ we obtain a cross section of the stability region in a subspace with $\underline{\Theta}$ coordinates. If the stability region in this cross section contains Ω, then $\underline{k}$ is a solution of the robustness problem. The search for such a $\underline{k}$ may be performed in discrete steps in $\underline{k}$.

ii) Similarly the stability region may also be cut by a subspace for constant $\underline{\Theta}$. The set of all stabilizing $\underline{k}$ for this particular value of $\underline{\Theta}$ is obtained. In the aircraft example the latter is the only possible approach because only a finite set of models for $\underline{\Theta}_1$, $\underline{\Theta}_2\ldots\underline{\Theta}_N$ are available. In this case the set of all simultaneously stabilizing $\underline{k}$ is the intersection of all stability regions in the subspaces for $\underline{\Theta} = \underline{\Theta}_1$, $\underline{\Theta} = \underline{\Theta}_2\ldots\underline{\Theta} = \underline{\Theta}_N$ projected into one K-space. This is the "multi-model approach".

The aircraft example suggests using the multi-model problem formulation with discretized $\underline{\Theta}$ also in other examples. For the bus we may first solve the multi-model problem for the corners A, B, C and D in the plant parameter region Ω of fig. 2. If a simultaneously stabilizing $\underline{k}$ has been found, then the approach i) can be used in order to analyze whether the total region Ω is contained in the stability region.

In this problem formulation, "stability" may be replaced by "nice stability" or "Γ-stability" as defined by a region Γ in the left half s-plane such that all closed-loop modes are rapidly decaying, well damped, and within a specified bandwidth. Fig. 5 shows an example of Γ.

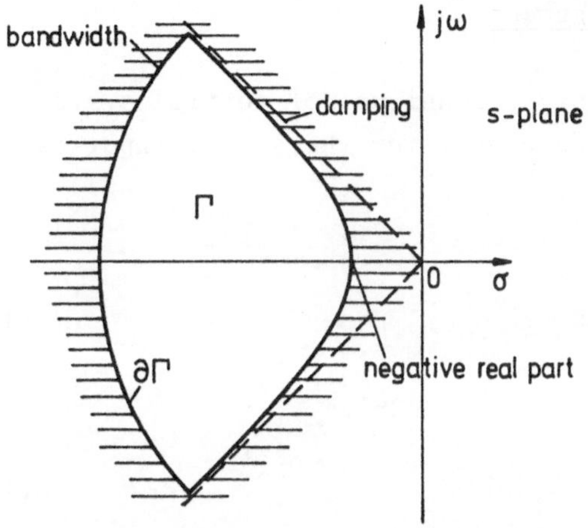

Fig. 5
A region of Γ-stability. The hyperbola boundary guarantees minimum damping and minimum negative real part of the eigenvalues. The bandwidth circle closes Γ.

Also other measures of performance P may be used to define admissible regions ($P < P_r$, see fig. 4).

The multi-model problem formulation is also used if failure situations must be considered in the design. In the aircraft example, a gyro or an accelerometer may fail. In an electric power network, a power station may be switched off from the network. Such possible operating conditions lead to additional discrete values of the plant parameters $\underline{\Theta}_j$ [28].

In frequency design methods the multi-model formulation may be viewed as a plant description by a family of Bode or Nyquist plots or logarithmic plots in the Nichols chart. This approach is suitable if the model is determined by frequency response measurements and if the closed-loop specifications are given in the frequency domain (e.g. bandwidth, gain and phase margins) [7].

In this paper it is assumed that a family of state-space models for the plant is given. This is true e.g. for many mechanical systems like vehicles or robots.

We assume also that essential closed-loop specifications can be formulated as a Γ-stability region in the s-plane for continuous systems or in the z-plane for discrete-time systems. In the next two sections the approaches of "simultaneous pole-region assignment" and "vectorial performance criteria" will be surveyed.

3. Simultaneous Pole-Region Assignment

Pole assignment is a standard design technique for control systems. For simplicity it will be discussed here only for the single-input case.

Given

$$\dot{\underline{x}} = \underline{A}\underline{x}+\underline{b}u, \quad (\underline{A},\underline{b}) \text{ controllable} \tag{16}$$

Find a state-feedback law

$$u = -\underline{k}'\underline{x}+r \tag{17}$$

such that the closed-loop system

$$\dot{\underline{x}} = (\underline{A}-\underline{b}\underline{k}')\underline{x}+\underline{b}r \tag{18}$$

has a specified set of eigenvalues $s_1, s_2 \ldots s_n$, i.e. the closed-loop characteristic polynomial is

$$P(s) = \det(s\underline{I}-\underline{A}+\underline{b}\underline{k}') = (s-s_1)(s-s_2)\ldots(s-s_n) \tag{19}$$

There are several techniques for solving this equation for $\underline{k}'$ [8]. The author [9] has derived the following solution

$$\underline{k}' = \underline{e}'P(\underline{A}) \tag{20}$$

where

$$\underline{e}' = [0 \ldots 0 \quad 1][\underline{b}, \underline{A}\underline{b} \ldots \underline{A}^{n-1}\underline{b}]^{-1} \tag{21}$$

Using the factorized form of P(s) in eq. (19),

$$\underline{k}' = \underline{e}'(\underline{A}-s_1\underline{I})(\underline{A}-s_2\underline{I})\ldots(\underline{A}-s_n\underline{I}) \tag{22}$$

The standard use of pole assignment is to specify the desired closed-loop eigenvalues $s_1, s_2 \ldots s_n$, and to calculate the required feedback gain vector $\underline{k}$. In pole region assignment only an admissible region Γ in s-plane is specified, see for example fig. 5, i.e. it is only required that

$$s_1, s_2 \ldots s_n \in \Gamma \tag{23}$$

This form of specification is more natural for the designer, because generally he does not know how to choose the s_i exactly, but he has a good idea about Γ. The solution of eq. (22) is then an admissible set K_Γ such that

$$s_1, s_2 \ldots s_n \in \Gamma \iff \underline{k} \in K_\Gamma \tag{24}$$

$\underline{k} = \underline{k}(s_i)$ is a continuous function. Therefore the boundary ∂K_Γ of the region K_Γ in K-space is obtained by mapping the boundary $\partial\Gamma$ of the region Γ in the s-plane. There are two kinds of boundaries, the real and the complex boundaries.

The real root boundary (rrb) at s = a maps into the set of $\underline{k}$' which places one eigenvalue, say s_1, at a and all other eigenvalues $s_2 \ldots s_n$ into Γ

$$\text{rrb} = \{\underline{k} \mid s_1 = a \quad , \quad s_2 \ldots s_n \in \Gamma\} \tag{25}$$

The complex root boundary (crb) in K space is the set of all $\underline{k}$ such that a complex conjugate pair of eigenvalues, say s_1, $s_2 = \bar{s}_1$, is located on the complex root boundary and all other eigenvalues $s_3 \ldots s_n$ are located in the region Γ.

$$\text{crb} = \{\underline{k} \mid (s_1, s_2 = \bar{s}_1) \in \partial\Gamma \ , \ s_3 \ldots s_n \in \Gamma\} \tag{26}$$

These boundaries of K_Γ are surfaces in K-space. Further properties of these boundaries and their calculation are described in [10], [11], [12], [13].

In comparison with pole assignment, pole region assignment offers more flexibility for simultaneous Γ-stabilization of a family of plant models $(\underline{A}_j, \underline{b}_j)$, j = 1,2...N, by one fixed gain vector $\underline{k}$. Each plant model $(\underline{A}_j, \underline{b}_j)$ gives rise to a corresponding admissible region $K_{\Gamma j}$ in K-space. The set of simultaneous Γ-stabilizers is the intersection

$$K_\Gamma = \bigcap_{j=1}^{N} K_{\Gamma j} \tag{27}$$

Fig. 6 illustrates eq. (27) for the case of two plant models.

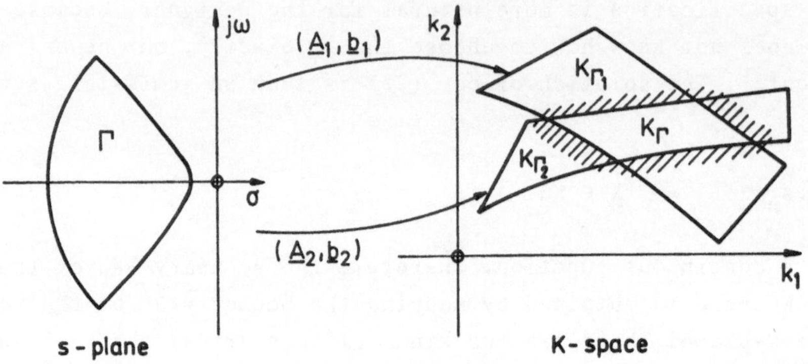

Fig. 6 $k \in K_{\Gamma_1}$ places all eigenvalues of $(\underline{A}_1-\underline{b}_1\underline{k}')$ into Γ,
$k \in K_{\Gamma_2}$ places all eigenvalues of $(\underline{A}_2-\underline{b}_2\underline{k}')$ into Γ.
The intersection $K_\Gamma = K_{\Gamma_1} \cap K_{\Gamma_2}$ describes the set of simultaneous Γ-stabilizers for the two plant models.

The design by simultaneous pole-region assignment may be carried out in different ways:

i) A graphical representation of K_Γ can be made in two-dimensional [11], [12] and three-dimensional [13] cross sections of K-space. The choice of the cross-section may be determined by the controller structure [5] or by the choice of invariance planes [3] such that some eigenvalues are not shifted in one design step. It may also be chosen such that the admissible area or volume in the subspace is maximized. The design point can be selected from the admissible set in consideration of additional design requirements [10], [12]

- small $||\underline{k}||$ in order to reduce $|u|$,
- a safety margin for $\underline{k}$ away from the boundaries of K_Γ for the case of implementation inaccuracies (e.g. quantization).
- robustness with respect to sensor failures,
- gain reduction margins.

This approach has been successfully applied to the aircraft [5], [6] and track-guided bus [3] problems.

ii) For a practically interesting class of regions Γ (including those with conic section bounds), Γ can be mapped onto the left half plane such that the problem is reduced to the stability problem [14]. The Hurwitz stability conditions lead to a set of linear and nonlinear inequalities in the elements of $\underline{k}$. Nonlinear optimization methods can be used to find a solution.

iii) The boundary $\partial\Gamma$ may be imbedded into a family of boundaries $\partial\Gamma_r$ with one parameter r. This family may be defined such that r is a measure for the degree of stability [12]. For sufficiently large r a simultaneous Γ_r-stabilizer always exists. Numerical algorithms may be developed such that Γ_r is contracted by reduction of r until the set of simultaneous Γ_r stabilizers reduces to one point in K-space.

4. Vectorial Performance Index

Design is a tradeoff between various competing objectives. Some typical design objectives have been formulated in the preceeding section in terms of an eigenvalue region. Other objectives are related to feedback gains and their margins; they have been used in the selection of a particular solution from the admissible set. Typically there are more design objectives which have not been considered so far. Examples are

- the deviation of the step response from a specified reference model response,
- good disturbance rejection,
- the actuator energy required for such responses,
- stability margins in the frequency domain.

For each of these requirements a performance index is defined such that its value is nonnegative and such that the smaller the value the better the requirement is met. For the multi-model problem $(\underline{A}_j, \underline{b}_j)$, j = 1,2...N, each performance index must be formed N times for the N members of the family. All these indices may be combined into a vector

$$\underline{g}(\underline{k}) = \begin{bmatrix} g_1(\underline{k}) \\ g_2(\underline{k}) \\ \vdots \\ g_M(\underline{k}) \end{bmatrix} \qquad (28)$$

A Pareto-optimal value of $\underline{g}$ is wanted. The definition of Pareto-optimality is illustrated by fig. 7 for the case of two indices g_1 and g_2.

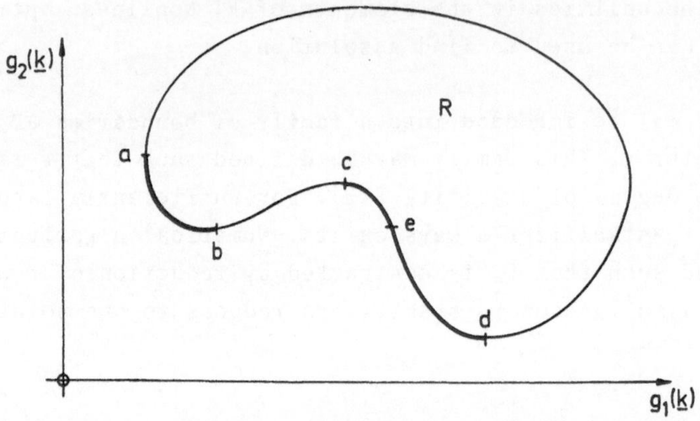

Fig. 7 On the definition of Pareto-optimal solutions.

The closed region R describes all possible values of the performance vector for admissible values of $\underline{k}$ (e.g. $\underline{k} \in K_\Gamma$ from pole region assignment, or all real $\underline{k}$). Now all $\underline{g}$-vectors are excluded for which, locally, both criteria g_1 and g_2 can be improved simultaneously. There remain some parts of the boundary, in the example of fig. 7 the parts ab and cd. In these portions one performance index can only be improved at the expense of the other one. These solutions are called Pareto-optimal. The portions ab and ed are globally Pareto-optimal, the portion ce is only locally Pareto-optimal. Ideally we try to find a globally optimal solution. But a procedure, which finds a locally optimal solution, is also a useful tool.

Suppose the designer knows the set of Pareto-optimal solutions. Then he has information on which tradeoffs are possible for a given plant and he can select the most desired solution from the Pareto-optimal set. Practically it is difficult to compute and visualize the set of Pareto-optimal solutions.

The design strategy of Kreisselmeier and Steinhauser [15] allows a systematic search for a desirable Pareto-optimal solution. During this interactive search the designer learns about the conflicts in design objectives and the possible tradeoffs. The idea is illustrated by fig.8.

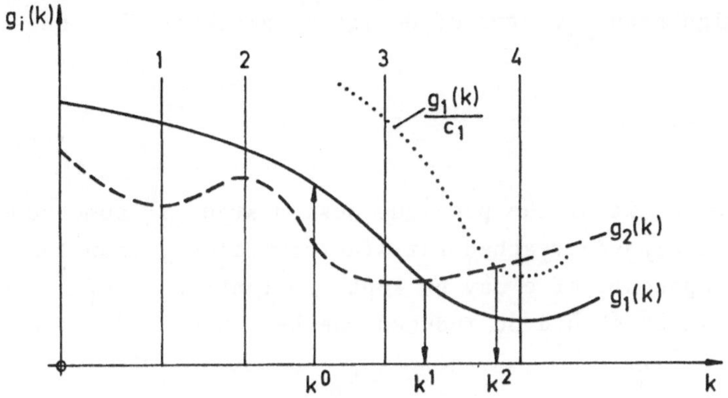

Fig. 8 Two performance indices g_1 and g_2 depending on only one controller parameter k.

Pareto-optimal solutions are located between the vertical lines 1 and 2 (local solutions) and between 3 and 4 (global solutions). Starting from an initial guess k^0 a Pareto-optimal solution k^1 is found by minimizing the maximal component of $\underline{g}$

$$\underline{g}^1 = \underline{g}(k^1) = \min_k \max\{g_1(k),\ g_2(k)\} \tag{29}$$

Now assume the designer does not like this solution. He wants to reduce g_1. He can achieve this by dividing $g_1(k)$ by a factor c_1, $0 < c_1 < 1$, i.e. he computes

$$\underline{g}^2 = g(k^2) = \min_k \max\left\{\frac{g_1(k)}{c_1},\quad g_2(k)\right\} \tag{30}$$

He obtains a different Pareto-optimal solution k^2. Each Pareto-optimal solution can be found by appropriate choice of c_1.

In general the design can be steered in a desired direction in a systematic way by the following procedure:

i) Choose $\underline{c}^1$ such that

$$\underline{g}(\underline{k}^0) < \underline{c}^1 \tag{31}$$

(The notation $\underline{a} < \underline{b}$ means $a_i \leq b_i$ for $i = 1,2...M$ and $\underline{a} \neq \underline{b}$).

ii) In the ν-th design step a vector of design parameters $\underline{c}^{\nu}$ is chosen such that

$$\underline{g}(\underline{k}^{\nu-1}) < \underline{c}^{\nu} < \underline{c}^{\nu-1} \tag{32}$$

$\underline{k}^{\nu-1}$ denotes the result of the previous design step. If some components of $\underline{g}$ have already reached satisfactory values, then the corresponding components of $\underline{c}$ may be kept constant: $c_i^{\nu} = c_i^{\nu-1}$. For components of $\underline{g}$ which should be reduced the best achieved value is chosen: $c_j^{\nu} = g_j(\underline{k}^{\nu-1})$.

iii) The solution of the min-max problem for finding $\underline{k}^{\nu}$ may be approximated by the scalar optimization problem [15], [16]

$$\min_{\underline{k}} \{\frac{1}{\rho}\, \ell n \sum_{i=1}^{L} \exp\,[\rho g_i(\underline{k})/c_i^{\nu}]\} \tag{33}$$

with sufficiently large ρ. Here an unconstrained optimization is used as a design tool instead of more complicated nonlinear programming techniques.

iv) The iteration terminates when

$$\underline{g}(\underline{k}^{\nu}) \approx \underline{c}^{\nu} \tag{34}$$

Then by construction the sequence of design vectors is monotonically decreasing, i.e.

$$\underline{g}(\underline{k}^{\nu}) < \underline{c}^{\nu} < \underline{c}^{\nu-1} <\ldots< \underline{c}^{1} \tag{35}$$

This design method is implemented in the computer program REMVG by Steinhauser. This program was successfully applied to the design of controllers for the aircraft control problem [16] for an electric power control system [17] and for the control of a cryogenic wind tunnel [18].

5. Parameter Uncertainty in Continuous Intervals

In the multi-model formulation we have represented the region Ω of the physical parameter vector $\underline{\Theta}$ by some typical or extremal points. Then the question arises: How can the stability of the closed loop be tested for all $\underline{\Theta} \in \Omega$. This can be done graphically if $\underline{\Theta}$ consists of only two plant parameters Θ_1 and Θ_2. Fig. 9 illustrates a two-dimensional region Ω.

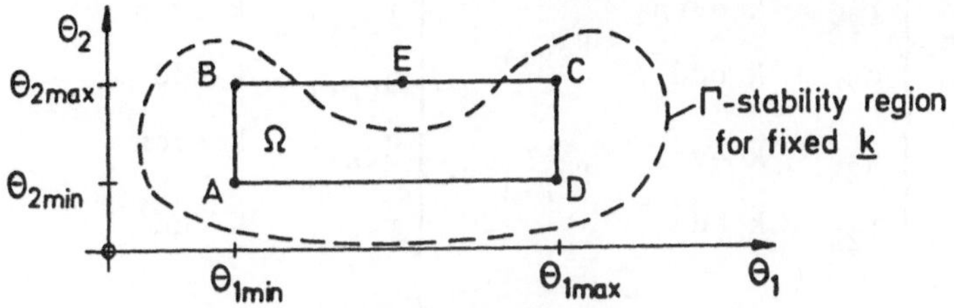

Fig. 9 Γ-stability was assured for the points A, B, C and D by a multi-model formulation. In the situation depicted the multi-model approach may be repeated for A, B, C, D and E.

Suppose $\underline{k}$ was determined by a multi-model approach for simultaneous Γ-stabilization of the models A, B, C and D. Then the Γ-stability boundary may be mapped into the $\underline{\Theta}$-plane, see for example the dotted line in fig. 9. By construction A, B, C and D must be located inside the Γ-stability region. But other points in Ω may be not Γ-stable as the figure shows. A practical engineering solution is to include the point E into the model family and to repeat the design. For a third plant parameter Θ_3, 3D computer graphics may be used or a "moving cross section", which can be moved in the direction of Θ_3. This approach and related software is not yet developed.

Also numerical techniques for investigating interval polynomials and interval matrices are not well developed. A basic result is Kharitonov's theorem [19]:

Assume that the coefficients of a polynomial

$$P(s) = p_o+p_1s+\ldots+p_{n-1}s^{n-1}+s^n \tag{36}$$

are unknown. The coefficients may take arbitrary values in the known intervals

$$q_i \le p_i \le r_i \tag{37}$$

A necessary and sufficient condition that P(s) is a Hurwitz polynomial is that the 2^n extremal polynomials with coefficients q_i and r_i are Hurwitz polynomials. This in turn is true if and only if the following four of these polynomials are Hurwitz

$$P_1(s) : p_{2k} = \begin{cases} r_{2k} & k \text{ even} \\ q_{2k} & k \text{ odd} \end{cases} \qquad p_{2k+1} = \begin{cases} r_{2k+1} & k \text{ even} \\ q_{2k+1} & k \text{ odd} \end{cases}$$

$$P_2(s) : p_{2k} = \begin{cases} q_{2k} & k \text{ even} \\ r_{2k} & k \text{ odd} \end{cases} \qquad p_{2k+1} = \begin{cases} q_{2k+1} & k \text{ even} \\ r_{2k-1} & k \text{ odd} \end{cases}$$

$$P_3(s) : p_{2k} = \begin{cases} q_{2k} & k \text{ even} \\ r_{2k} & k \text{ odd} \end{cases} \qquad p_{2k+1} = \begin{cases} r_{2k+1} & k \text{ even} \\ q_{2k+1} & k \text{ odd} \end{cases} \tag{38}$$

$$P_4(s) : p_{2k} = \begin{cases} r_{2k} & k \text{ even} \\ q_{2k} & k \text{ odd} \end{cases} \qquad p_{2k+1} = \begin{cases} q_{2k+1} & k \text{ even} \\ r_{2k+1} & k \text{ odd} \end{cases}$$

Example:

$$P(s) = p_0+p_1s+p_2s^2+p_3s^3+s^4 \tag{39}$$

with given intervals $q_0 \le p_0 \le r_0$, $q_1 \le p_1 \le r_1$, $q_2 \le p_2 \le r_2$, $q_3 \le p_3 \le r_3$ is a Hurwitz polynomial if and only if this is true for the polynomials

$$\begin{aligned} P_1(s) &= r_0 + r_1s + q_2s^2 + q_3s^3 + s^4 \\ P_2(s) &= q_0 + q_1s + r_2s^2 + r_3s^3 + s^4 \\ P_3(s) &= q_0 + r_1s + r_2s^2 + q_3s^3 + s^4 \\ P_4(s) &= r_0 + q_1s + q_2s^2 + r_3s^3 + s^4 \end{aligned} \tag{40}$$

Unfortunately the corresponding result does not hold for interval matrices. Barmish [20] gave the following counterexample:

$$\underline{A}(\Theta) = \begin{bmatrix} -\Theta & -12.06 & -0.06 \\ -0.25 & 0 & 1 \\ 0.25 & -4 & -1 \end{bmatrix} \tag{41}$$

$$P(s,\Theta) = \det[s\underline{I}-\underline{A}(\Theta)] = (4\Theta+0.06) + (1+\Theta)s + (1+\Theta)s^2 + s^3 \tag{42}$$

$$Q(s) = P(s,0.5) = 2.06 + 1.5s + 1.5s^2 + s^3$$
$$R(s) = P(s,1.5) = 6.06 + 2.5s + 2.5s^2 + s^3$$

Both are Hurwitz polynomials, i.e. both $\underline{A}(0.5)$ and $\underline{A}(1.5)$ are stable. But the intermediate matrix $\underline{A}(1)$ is unstable.

For the characteristic polynomial of an interval matrix, eq. (38) provides only a sufficient condition. It is not necessary because the polynomial coefficients are not independent, i.e. some coefficient sets admissible in eq. (37) cannot occur. This may be illustrated by a slight modification of the example (41). Change the element a_{13} from -0.06 to +0.08, then

$$P(s,\Theta) = (4\Theta-0.08) + (0.98+\Theta)s + (1+\Theta)s^2 + s^3 \tag{43}$$

This is a Hurwitz polynomial for all $\Theta > 0.02$ and therefore also in the interval $0.5 \leq \Theta \leq 1.5$. The four polynomials of eq. (38) are

$$\begin{aligned} P_1(s) &= 5.92 + 2.48s + 1.5s^2 + s^3 \\ P_2(s) &= 1.92 + 1.48s + 2.5s^2 + s^3 \\ P_3(s) &= 1.92 + 2.48s + 2.5s^2 + s^3 \\ P_4(s) &= 5.92 + 1.48s + 1.5s^2 + s^3 \end{aligned} \tag{44}$$

The polynomials $P_1(s)$ and $P_4(s)$ are unstable, thus independent variations of p_0, p_1, p_2 between the extremal values of eq. (43) for $\Theta = 0.5$ and $\Theta = 1.5$ are not admissible. This example shows again, that it is not useful to imbed a problem into a much larger class of problems for which a solution exists. The knowledge about the structure of the parameter uncertainty should be used in order to get tight bounds.

Note also that Kharitonov's result does not generalize to other Γ-stability regions. A Lyapunov approach to stability of interval matrices was made by Kiendl [21]. A sufficient condition is derived from the negative definiteness of the derivative of the Lyapunov function for all extremal combinations of the matrix elements. Again independent uncertainties of the matrix elements are assumed.

Zeheb and Hertz [22] study the characteristic equation (15) along the boundary $\partial\Gamma$ of the nice stability region in s-plane.

$$P(s,\underline{\Theta},\underline{k})\Big|_{s\in\partial\Gamma} = 0 \tag{45}$$

Admissible intervals for plant parameters $\underline{\Theta}_i$ and feedback gains k_i are calculated successively.

Klickow and Franke [33] combine various necessary conditions for the enclosure of eigenvalues in the complex plane by Gershgorin circles.

6. Existence and Dynamic Order of Simultaneous Γ-Stabilizers

Only few results are available on the question of the structure and dynamic order of controllers for a family of plant models. From an engineering point of view the recommended procedure is to assume a simple controller structure and try to find feasible parameters in this structure, for example by simultaneous pole region assignment or optimization of vectorial performance indices. If a Γ-stabilizing solution does not exist, then the controller structure may be extended. This practical approach has led to good engineering solutions in various applications.

From a theoretical point of view it remains unsatisfactory, that a priori assumptions on the controller structure must be made and no general conditions on the existence of fixed gain controllers are known. Even the most elementary problem of simultaneous stabilization of two plant models by the same controller is not yet fully understood. Conditions for simultaneous stabilizability were studied by Saeks and Murray [23] and Vidyasagar and Viswanadham [24] for the linear case and by Desoer and Lin [25] for the nonlinear case. But these results do not yet help the designer to choose the structure and dynamic order of a simultaneous stabilizer or Γ-stabilizer. Olbrot [26] proposed a time-varying controller for simultaneous stabilization of a family of N plant models. It switches cyclically between N deadbeat controllers. The controller which corresponds to the actual plant will bring the state to zero.

For fixed-gain controllers there is no general upper limit on the order m of a dynamic state-feedback controller for simultaneous stabilization of two discrete-time plants. This was shown by the author [27] by an example with a parameter a. For any chosen m there is a value of a for which simultaneous stabilization is impossible but for the same value of a, m can be chosen large enough such that simultaneous stabilization becomes possible. Thus the required controller order grows monotonically with the parameter a and there is no upper limit on the integer m.

The introduction of dynamics in the state-feedback controller does not help to make the simultaneous stabilization problem solvable, if there is a conflict of the real root conditions [27]. For discrete-time systems the real root condition for $z = 1$ is for example $P(1) > 0$. If there does not exist a state-feedback vector $\underline{k}'$ which simultaneously satisfies the real root conditions

$$P(1) = \det(\underline{I}-\underline{A}_j+\underline{b}_j\underline{k}') > 0 \qquad (46)$$

for all j, then the discrete-time plant family $(\underline{A}_j,\underline{b}_j)$ cannot be stabilized by a linear controller of any order m. The corresponding result holds for the other real root condition at z = -1, i.e. $(-1)^n P_j(-1) > 0$ for all j.

7. Conclusions

For mechanical systems like vehicles and robots it is frequently possible to describe the plant by a family of models $(\underline{A}_j,\underline{b}_j)$, j = 1,2...N, for different operating or failure conditions. Methods for the design of a common fixed-gain controller for this plant family have been developed. A controller structure and order are assumed and feasible parameters are sought. It would be desirable to have a theory which helps the designer to find controller structure and order systematically and also to identify situations, where no simultaneous stabilizer exists. So far the theory is of little help in this problem.

The investigation of diverse practical examples has shown that also large parameter variations can be accomodated by a fixed-gain controller. The previous lack of methods for design of robust fixed-gain controllers has led to more complicated gain-scheduling solutions, for example for flight control problems. These can be simplified now. Also in cases where a fixed gain controller cannot achieve the same performance as a gain scheduling or adaptive system, it may be used at the lowest hierarchical level such that the higher level variable gain system sees a well-behaved (e.g. Γ-stable) controlled plant instead of the open-loop plant with instability and larger uncertainty. This approach also reduces the redundancy requirements which must be satisfied only at the relatively simple lowest hierarchical level. In systems with a human operator it is possible for him to take over manual control if a higher level automatic system fails, because he has to deal with a Γ-stabilized plant. Also the problem of sensor failure detection is simplified by a robust stabilization at the lowest hierarchical level [28].

Two design methods for robust controllers have been developed at DFVLR: Simultaneous Γ-stabilization by the parameter space method and optimization of vector performance criteria. Both of them will profit from the rapid technical development in computer graphics including color and 3D representations. Graphic representations allow the designer to visualize the conflicts and possibilities so that he can decide interactively on tradeoffs during the design process.

Both methods are well suited for the multi-model formulation of robust control. The verification of the results for intermediate constant plant parameter values is possible but still tedious. If the real plant is nonlinear or time-varying, then a simulation study and fine tuning of controller parameters is necessary anyway and may replace the study of interval matrices or polynomials for the "frozen" local description. The multi-model description is certainly not perfect for nonlinear or time-varying systems, but it is an important step to a more realistic consideration of plant uncertainty than it is possible with the classical sensitivity methods.

8. Literature

[1] Christ, H., Darenberg, W., Panik, F., Weidemann, W.: Automatic track control of vehicles. 5th VSD and 2nd IUTAM Symposium, Vienna, Sept. 1977.

[2] Darenberg, W., Gipser, M., Türk, S.: Probleme der robusten Regelung aus dem Bereich der Kraftfahrzeugforschung. Interkama Congress, Düsseldorf, Oct. 1983., Springer, Berlin 1983, 319-330.

[3] Ackermann, J., Türk, S.: A common controller for a family of plant models. 21st IEEE Conference on Decision and Control, Orlando, Dec. 1982, 240-244.

[4] Berger, R.L., Hess, J.R., Anderson, D.C.: Compatibility of maneuver load control and relaxed static stability applied to military aircraft. AFFDL-TR-73-33, 1973.

[5] Franklin, S.N., Ackermann, J.: Robust flight control: a design example. AIAA Journal of Guidance and Control 1981, vol. 4, 597-605.

[6] Gruebel, G., Joos, D., Kaesbauer, D., Hillgren, R.: Robust back up stabilization for artificial stability aircraft. 14th ICAS Congress, Toulouse, Sept. 84.

[7] Horowitz, I.: Quantitative synthesis of uncertain multiple input-output feedback systems. Int. J. Control 1979, vol. 30, 81-106.

[8] Kailath, T.: Linear systems. Englewood Cliffs, N.J.: Prentice Hall 1980.

[9] Ackermann, J.: Der Entwurf linearer Regelungssysteme im Zustandsraum. Regelungstechnik 1972, vol. 20, 297-300.

[10] Ackermann, J.: Parameter space design of robust control systems. IEEE Trans. Aut. Control 1980, vol. 25, 1058-1980.

[11] Ackermann, J., Kaesbauer, D.: D-decomposition in the space of feedback gains for arbitrary pole regions. 8th IFAC Congress, Kyoto, Aug. 1981, vol. IV, 12-17.

[12] Ackermann, J.: Abtastregelung. Berlin, Springer 1983. English version "Sampled-data control systems", to appear 1985.

[13] Putz, P.: Algorithms for modeling and display of 2D and 3D projections of parameter space regions for siso control. Center of Interactive Computer Graphics, Rensselaer Polytechnic Institute, Troy, N.Y., Technical Report No. 84001.

[14] Sondergeld, K.P.: A generalization of the Routh-Hurwitz stability criteria and an application to a problem in robust controller design. IEEE Trans. Aut. Control 1983, vol. 28, 965-970.

[15] Kreisselmeier, G., Steinhauser, R.: Systematische Auslegung von Reglern durch Optimierung eines vektoriellen Gütekriteriums. Regelungstechnik 1979, vol. 27, 76-79.

[16] Kreisselmeier, G., Steinhauser, R.: Application of vector performance optimization to a robust control loop design for a fighter aircraft. Int. J. Control 1983, vol. 37, 251-284.

[17] Björnsson, B., Cuno, B., Handschin, E., Voss, J.: Auslegung robuster Regelsysteme in der elektrischen Energieversorgung. Interkama Congress, Düsseldorf, Oct. 1983, Springer, Berlin 1983, 308-318.

[18] Steinhauser, R.: Reglerentwurf für einen Tieftemperatur-Windkanal mittels Gütevektor-Optimierung. Dissertation TU Karlsruhe 1984, and DFVLR-Forschungsbericht, to appear 1985.

[19] Kharitonov, V.L.: Asymptotic stability of an equilibrium position of a family of systems of linear differential equations. Differentsial'nye Uraveniya 1978, vol. 14, 2080-2088. English translation Plenum Publishing Corporation 1979, 1483-1485.

[20] Barmish, B.R., Hollot, C.V.: Counter-example to a recent result on the stability of interval matrices by S. Bialas. Int. J. Control 1984, vol. 39, 1103-1104.

[21] Kiendl, H.: Stabilitätsnachweis für den Multimodellansatz mit einem Kontinuum von möglichen Parameterwerten. Fifth GMR Workshop on Robust Control, Interlaken, Oct. 1984.

[22] Zeheb, E., Hertz, D.: Robust control of the characteristic values of systems with possible parameter variations. Int. J. Control 1984, vol. 40, 81-96.

[23] Saeks, R., Murray, J.: Fractional representation, algebraic geometry, and the simultaneous stabilization problem. IEEE Trans. Aut. Control 1982, vol. 27, 895-903.

[24] Vidyasagar, M., Viswanadham, N.: Algebraic design techniques for reliable stabilization. IEEE Trans. Aut. Control 1982, vol. 27, 1085-1095.

[25] Desoer, C.A., Lin, C.A.: Simultaneous stabilization of nonlinear systems. IEEE Trans. Aut. Control 1984, vol. 29, 455-457.

[26] Olbrot, A.W.: A simple method of robust stabilization of linear uncertain systems. Proc. Conf. Measurement and Control, Athens, Sept. 1983.

[27] Ackermann, J.: Simultaneous stabilization of two plant models. Proc. Pre-IFAC Meeting on Current Trends in Control, Dubrovnik-Cavtat, June 1984.

[28] Ackermann, J.: Robustness against sensor failures. Automatica 1984, vol. 20, 31-38.

[29] Kwakernaak, H.: Uncertainty models and the design of robust control systems. This volume.

[30] Åström, K.J.: Adaptive control - a way to deal with uncertainty. This volume.

[31] Tsypkin, Ya.: Optimality in adaptive control systems. This volume.

[32] Gruebel, G.: Uncertanty and control - some activities at DFVLR. This volume.

[33] Klickow, H.H., Franke, D.: Eigenwerteinschließung für kontinuierliche und diskrete Regelungen bei Berücksichtigung von Parameterstörungen. XIX Regelungstechnisches Kolloquium, Boppard, Feb. 1985.

Adaptive Control
– A Way to Deal with Uncertainty

Karl Johan Aström

Department of Automatic Control
Lund Institute of Technology
Box 118, S-221 00 Lund, Sweden

Abstract. This paper approaches the uncertainty problem from the point of view of adaptive control. The uncertainty is reduced by continuous monitoring of the response of the system to the control actions and appropriate modifications of the control law. It is shown that this approach makes it possible to deal with uncertainties that cannot be handled by high gain robust feedback control.

1. INTRODUCTION

The problem of reducing the consequences of uncertainty has always been a central issue in the field of automatic control. Black's invention of the feedback amplifier was motivated by the desire to make electronic circuits less sensitive to the variability of electronic tubes. The development of modern instrumentation technology has similarly made use of feedback in the form of the force balance principle, to make high quality instruments which are only moderately sensitive to variations in their components.

Feedback by itself has the ability to reduce the sensitivity of a closed loop system to plant uncertainties. Although this was one of the original motivations for introducing feedback, the idea was kept in the background during the intensive development of modern control theory. Lately the problem has received renewed interest. It is now a very active research field and several new schemes for robust control have recently been developed. Such shcemes typically result in constant gain feedback controls, which are insensitive to variations in plant dynamics. The possibilities and limitations of constant gain feedback are treated in Section 2. The purpose is to find out when a constant gain feedback can be designed to overcome uncertainty in process dynamics and when it can not.

An integrator where the sign of the gain is not known is a simple example which can not be handled by constant gain feedback. This example will be used as an illustration throughout the paper.

The main goal of the paper is to approach elimination of uncertainties from the point of view of adaptive control. When the plant uncertainties are such taht they can not be handled by a constant gain robust control law it is natural to try to reduce the uncertainties by

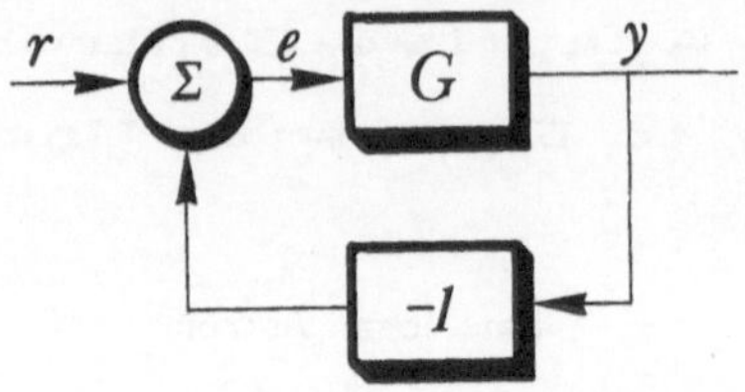

Figure 1. Simple feedback system

experimentation and parameter estimation. Auto-tuning is a simple technique, which has the attractive feature that an appropriate input signal to the process is generated automatically. The method has the additional benefit that parameter estimation and control design are extremely simple to do. This is discussed in Section 3.

Auto-tuning is an intermittent procedure. The regulator has a special tuning mode, which is invoked on the request of an operator or based on some automatic diagnosis. Adaptive control is a method which allows continuous reduction of the uncertainties. An adaptive regulator will continuously monitor the systems response to the control actions and modify the regulator appropriately. The charateristics of such control schemes are discussed in Section 4. Two categories of adaptive control laws, direct and indirect, are discussed in some detail. Some theoretical results on the stability of adaptive control systems are reviewed in Section 5. It is found that the standard assumptions used to prove the stability of direct adaptive control schemes are such that robust high gaiin linear control could equally well be applied. Adaptive controllers are nonlinear feedback systems. There are other types of nonlinear feedback systems, which also can deal with uncertainties. One type is called universal stabilizer. Such a system is briefly discussed in Section 6. Its capability of dealing with an integrator with unknown gain is demonstrated.

Stochastic control theory is a general method of dealing with uncertainties. In Section 7 it is shown how adaptive control laws can be derived from stochastic control theory. The example with the integrator having unknown gain is worked out in some detail.

2. LIMITATIONS OF CONSTANT GAIN FEEDBACK

Conventional feedback can deal with uncertainty in the form of disturbances and modeling errors. Before discussing other techniques for dealing with uncertainty it is useful to understand the possibilities and limitations of constant gain feedback. For this purpose consider the simple feedback system shown in Figure 1. Let G_0 be the nominal loop transfer function. Assume that true loop transfer function is $G = G_0(1+L)$ due to model uncertainties. Notice that $1+L$ is the ratio between the true and nominal transfer functions.

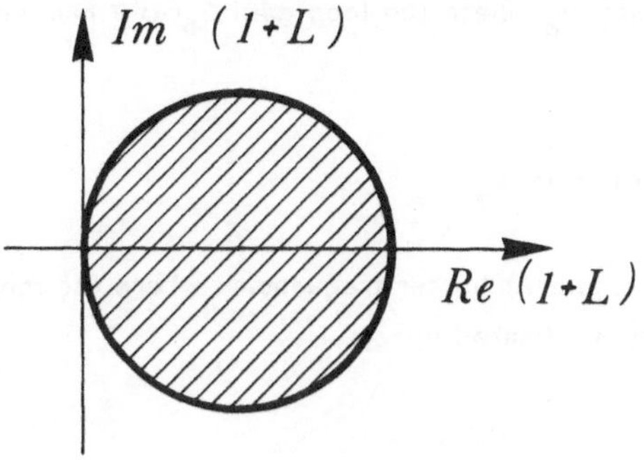

Figure 2. The ratio between the true transfer function G(s) and the nominal transfer function $G_0(s)$ must be in the shaded region for those frequencies where $G_0(i\omega)$ is large.

The effect of uncertainties on the stability of the closed loop system will first be discussed. The closed loop poles are the zeros of the equation

$$1 + G_0(s) + G_0(s)L(s) = 0$$

Provided that the nominal system is stable it follows from Rouche's theorem that the uncertainties will not cause instability provided that

$$|L(s)| \le \left|\frac{1+G_0(s)}{G_0(s)}\right| \tag{2.1}$$

on a contour which encloses the left half plane. The consequences of this inequality will now be discussed. For large loop gains (2.1) reduces to

$$|L(s)| \le 1$$

This means that the relative uncertainty 1+L must be in the shaded area in Figure 2. It follows from Figure 2 that if the uncertainty in the phase of the open loop system is less than φ in magnitude i.e.

$$|\arg(1+L)| \le \varphi$$

then the closed loop system is stable provided that the magnitude of the relative uncertainty satisfies

$$0 \le |1+L| \le 2\cos\varphi$$

For these frequencies where the loop gain is high it is thus necessary that the phase uncertainty is less than 90°.

At the crossover frequency ω_c where the loop gain $G_0(i\omega_c)$ has unit magnitude equation (2.1) reduces to

$$|L(i\omega_c)| \le \sqrt{2(1-\cos\varphi_m)} \tag{2.2}$$

where φ_m is the phase margin. At higher frequencies where the loop gain is less than one the inequality (2.1) can be approximated by

$$|L(s) \le \left|\frac{1}{G_0(s)}\right|$$

This means that large uncertainties can be treated where the loop gain is significantly less than one.

Stability is only a necessary requirement. To investigate the effect of uncertainty on the performance of the closed loop system consider the transfer function from the command signal to the output i.e.

$$G_0 = \frac{G}{1+G} = \frac{G_0+G_0L}{1+G_0+G_0L} = \frac{G_0}{1+G_0} \cdot \frac{1+L}{\dfrac{G_0}{1+G_0L}}$$

The error in the closed loop transfer function is thus

$$L_c = \frac{L}{1+G_0+G_0L} \tag{2.3}$$

This error can be made small either by having a small open loop uncertainty (L) or by having a high loop gain (G_0).

Equations (2.1) and (2.3) give the essence of high gain robust control. The open open loop gain G_0 can be made large for those frequencies where the phase uncertainty is less than 90 degrees. At those frequencies the closed loop transfer function can be made arbitrarily close to the specifications by choosing the gain sufficiently large. For those frequencies where the uncertainty in the phase shift is larger than 90° the total loop gain must be made smaller than one in order to maintain robustness. At the crossover frequency where the loop gain has unit magnitude the allowable phase uncertainty is given by (2.2). The allowable uncertainty depends critically on the phase margin φ_m. Assuming for example that it is desired to have an error in the closed loop transfer function of at most 10% of the crossover frequency. The allowable phase margin is given in Table 1.

Table 1 - Maximum error in the open loop transfer function which give at most 10 % error of the closed loop transfer function at the crossover frequency.

φ_m	10	20	30	45	60
max |L|	0.077	0.034	0.052	0.076	0.100

Design techniques which can deal with uncertainty are given in Gutman (1979), Horowitz (1963), Horowitz and Sidi (1973), Leitmann (1980,1983), Kwakernaak (1985), Grübel (1985). A discussion of the multivariable case is given by Doyle and Stein (1981).

It is clear from the discussion above that in order to use robust high gain control it is necessary that the transfer function of the plant has a phase uncertainty less than 90° for some frequencies. Some examples which illustrate the limitations of high gain robust control will now be discussed.

<u>Example 2.1</u> - Time Delays

Consider a linear plant where the major uncertainty is due to variations in the time delay. Assume that the time delay varies between T_{min} and T_{max}. Furthermore assume that it is required to keep the variations in the phase margin less than 20°. It then follows that the cross-over frequency ω_c must satisfy

$$\omega_c \leq \frac{0.35}{T_{max} - T_{min}}$$

The uncertainty in the time delay thus induces an upper bound to the achievable cross-over frequency. □

<u>Example 2.2</u> - Mechanical Resonances

Mechanical resonances are associated with transfer functions of the type

$$G(s) = \frac{\omega_0^2}{s^2 + 2\xi\omega_0 s + \omega_0^2}$$

where the damping normally is very small. The phase of G changes rapidly from 0 to -180° around ω_0. The gain also changes rapidly around ω_0. It increases from one to approximately $1/2\,\xi$ and it increases as ω_0^2/ω^2 with increasing ω. Variations in ξ and ω will thus give substantial phase uncertainty. To achieve robust linear control it is then necessary to make sure that the loop gain is low around ω_0. This is typically achieved by a notch filter. □

<u>Example 2.3</u> - Integrator Whose Sign is Unknown

An integrator whose sign is not known has either a phase lag of 90° or 270°. Such a system can not be controlled using high gain robust control. □

3. AUTO-TUNING

When the uncertainty is such that robust high gain feedback cannot be applied it is natural to try to reduce the uncertainty by experimentation. Auto-tuning is a methodology for doing this automatically. The principles are straightforward. A model of the process dynamics is determined by making an identification experiment where an input signal is generated and applied to the process. The dynamics of the process is then determined from the results of the experiment. The controller parameters are then obtained from some design procedure. Since the signal generation, the identification and the design can be made in many different ways there are many possible tuning procedures of this kind.

Auto-tuning is also useful in another context. There are cases where is is much easier to apply an auto-tuner than to design a robust high gain controller Simple regulators with two or three parameters can be tuned manually if there is not too much interaction between adjustments of different parameters. Manual tuning is, however, not possible for more complex regulators. Traditionally tuning of such regulators have followed the route of modeling or identification and regulator design. This is often a time-consuming and costly procedure which can only be applied to important loops or to systems which are made in large quantities.

Most adaptive techniques can be used to provide automatic tuning. In such applications the adaptation loop is simply switched on and perturbation signals may be added. The adaptive regulator is run until the performance is satisfactory. The adaptation loop is then disconnected and the system is left running with fixed regulator parameters. Below we will discuss a specific auto-tuner which requires very little prior information and also has the interesting property that it generates an appropriate test signal automatically. This is discussed further in Åström and Hägglund (1984a). A nice feature of the technique described below is that an input signal is generated automatically and that the parameter estimation and the control design are very simple. The input signal generated is automatically tuned to the characteristics of the plant. It will have its energy concentrated around the frequencies where the plant has phase lag of 180°.

The Basic Idea

A wide class of process control problems can be described in terms of the intersection of the Nyquist curve of the open loop system with the negative real axis, which is traditionally

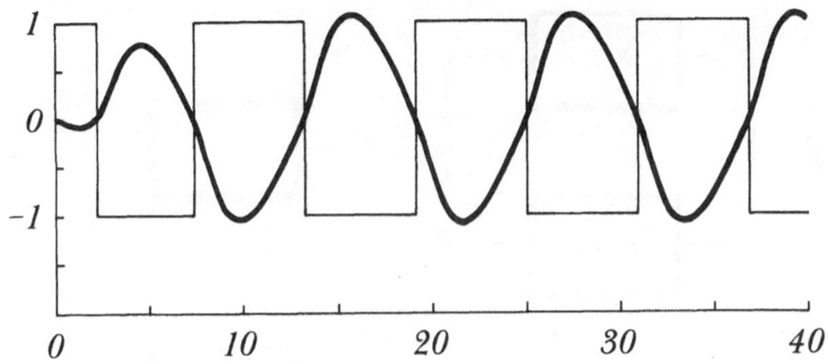

Figure 3. Input and output signals for a linear system under relay control. The system has the transfer function $G(s) = 0.5(1-s)/s(s+1)(s+1)$.

described in terms of the critical gain k_c and the critical period T_c. A method for determining these parameters was described in Ziegler and Nichols (1943). It is done as follows: A proportional regulator is connected to the system. The gain is gradually increased until an oscillation is obtained. The gain when this occurs is the critical gain and the frequency of the oscillation is the critical frequency. It is, however, difficult to perform this experiment in such a way that the amplitude of the oscillation is kept under control.

Relay feedback is an alternative to the manual tuning procedure. If the process is connected to a feedback loop there will be an oscillation as is shown in Figure 3. The period of the oscillation is approximately the critical period. The process gain at the corresponding frequency is approximately given by

$$G_p\left(i\ \frac{2\pi}{T_c}\right) \doteq -\ \frac{a\pi}{4d} \qquad (3.1)$$

where d is the relay amplitude and a is the amplitude of the oscillation.

A simple relay control experiment thus gives the desired information about the process. This method has the advantage that it is easy to control the amplitude of the limit cycle by an appropriate choice of the relay amplitude. A simple feedback from the output amplitude to the relay amplitude makes it possible to keep the output amplitude fixed during the experiment. Notice also that an input signal which is almost optimal for the estimation problem is generated automatically. This ensures that the critical point can be determined accurately.

When the critical point on the Nyquist curve is known, it is straightforward to apply the classical Ziegler-Nichols design methods. It is also possible to devise many other design schemes that are based on the knowledge of one point on the Nyquist curve. The procedure can be modified to determine other points on the Nyquist curve. An integrator may be connected in the loop after the relay to obtain the point where the Nyquist curve intersects

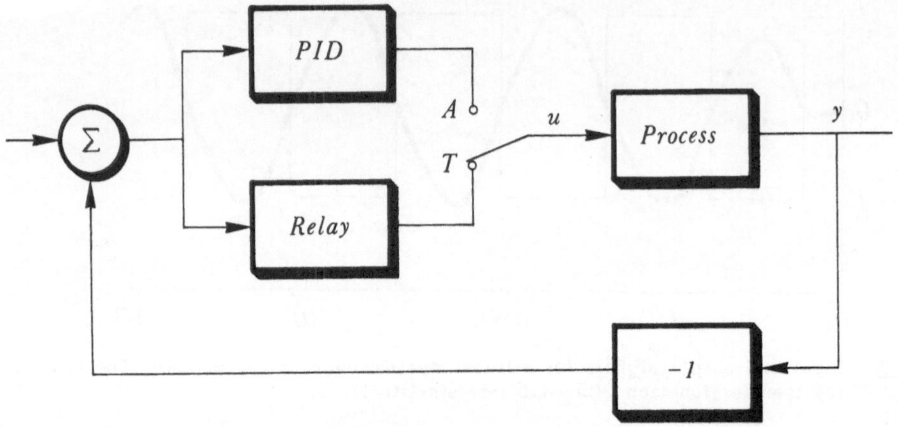

Figure 4. Block diagram of an auto-tuner. The system operates as a relay controller in the tuning mode (T) and as an ordinary PID regulator in the automatic control mode (A).

the negative imaginary axis. New design methods, which are based on such experiments, are described in Åström and Hägglund (1984b).

Methods for automatic determination of the frequency and the amplitude of the oscillation will be given to complete the description of the estimation method. The period of an oscillation can be determined by measuring the times between zero-crossings. The amplitude may be determined by measuring the peak-to-peak values of the output. These estimation methods are easy to implement because they are based on counting and comparisons only. More elaborate estimation schemes like least squares estimation and extended Kalman filtering may also be used to determine the amplitude and the frequency of the limit cycle oscillation. Simulations and experiments on industrial processes have indicated that little is gained in practice by using more sophisticated methods for determining the amplitude and the period.

A block diagram of a control system with auto-tuning is shown in Figure 4. The system can operate in two modes. In the tuning mode a relay feedback is generated as was discussed above. When a stable limit cycle is established its amplitude and period are determined as described above and the system is then switched to the automatic control mode where a conventional PID control law is used.

Practical Aspects

There are several practical problems which must be solved in order to implement an auto-tuner. It is e.g. necessary to account for measurement noise, level adjustment, saturation of actuators and automatic adjustment of the amplitude of the oscillation. It may be advantageous to use other nonlinearities than the pure relay. A relay with hysteresis gives a system which is less sensitive to measurement noise.

Measurement noise may give errors in detection of peaks and zero crossings. A hysteresis in the relay is a simple way to reduce the influence of measurement noise. Filtering is another possibility. The estimation schemes based on least squares and extended Kalman filtering can be made less sensitive to noise. Simple detection of peaks and zero crossings in combination with an hysteresis in the relay has worked very well in practice. See e.g. Åström (1982).

The process output may be far from the desired equilibrium condition when the regulator is switched on. In such cases it would be desirable to have the system reach its equilibrium automatically. For a process with finite low-frequency gain there is no guarantee that the desired steady state will be achieved with relay control unless the relay amplitude is sufficiently large. To guarantee that the output actually reaches the reference value, it may be necessary to introduce manual or automatic reset.

It is also desirable to adjust the relay amplitude automatically. A reasonable approach is to require that the oscillation is a given percentage of the admissible swing in the output signal.

Auto-Tuning with Learning

Auto-tuning is a simple way to reduce uncertainty by experimentation. In many cases the characteristics of a process may depend on the operating conditions. If it is possible to measure some variable which correlates well with the changing process dynamics it is possible to obtain a system with interesting characteristics by combining the auto-tuner with a table look-up function. When the operating condition changes a new tuning is performed on demand from the operator. The resulting parameters are stored in a table together with the variable which characterizes the operating condition. When the process has been operated over a range covering the operating conditions the regulator parameters can be obtained from the table. A new tuning is then required only when other conditions change. A system of this type is semi-automatic because the decision to tune rests with the operator. The system will, however, continue to reduce the plant uncertainty.

4. ADAPTIVE CONTROL

Adaptive control is another way to deal with uncertainties. A block-diagram of a typical adaptive regulator is shown in Figure 5. The system can be thought of as composed of two loops. The inner loop consists of the process and an ordinary linear feedback regulator. The parameters of the regulator are adjusted by the outer loop, which is composed of a recursive parameter estimator and a design calculation. To obtain good estimates it may also be

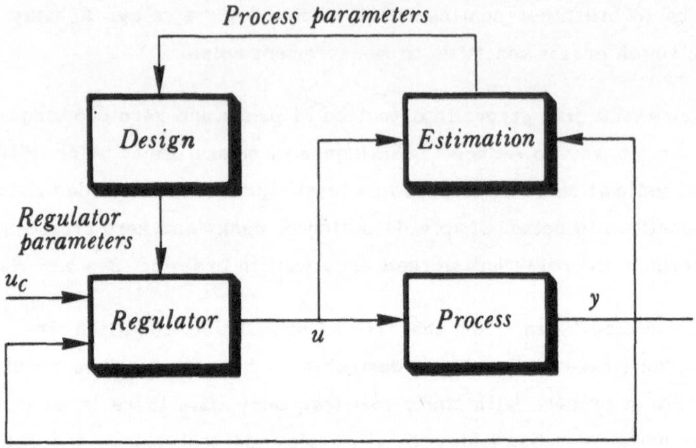

Figure 5. Block diagram of an adaptive regulator

necessary to introduce perturbation signals. This function is omitted from the figure for simplicity. Notice that the system may be reviewed as an automation of process modeling and design where the process model and the control design is updated at each sampling period.

The block labeled "regulator design" in Figure 5 represents an on-line solution to a design problem for a system with known parameters. This underlying design problem can be solved in many different ways. Design methods based on on phase- and amplitude margins, pole-placement, minimum variance control, linear quadratic gaussian control and other optimization methods have been considered, see Aström (1983). Robust design techniques can of course also be used.

The adaptive regulator also contains a recursive parameter estimator. Many different estimation schemes have been used, for example stochastic approximation, least squares, extended and generalized least squares, instrumental variables, extended Kalman filtering and the maximum likelihood method.

The adaptive regulator shown in Figure 5 is called indirect or explicit because the regulator parameters are updated indirectly via estimation of an explicit process model. It is sometimes possible to reparameterize the process so that it can be expressed in terms of the regulator parameters. This gives a significant simplification of the algorithm because the design calculations are eliminated. In terms of Figure 5 the block labelled design calculations disappears and the regulator parameters are updated directly. The scheme is then called a direct scheme. Direct and indirect adaptive regulators have different properties which is illustrated by an example.

Example 3.1

Consider the discrete time system described by

$$y(t+1) + \underline{ay(t)} = \underline{bu(t)} + e(t+1) + \underline{ce(t)} \qquad t=\ldots-1,0,1,\ldots \tag{4.1}$$

where $\{e(t)\}$ is a sequence of zero-mean uncorrelated random variables. If the parameters a, b and c are known the proportional feedback

$$u(t) = -\theta y(t) = \frac{a-c}{b} y(t) \tag{4.2}$$

minimizes the variance of the output. The output then becomes

$$y(t) = e(t) \tag{4.3}$$

This can be concluded from the following argument. Consider the situation at time t. The variable $e(t+1)$ is independent of $y(t)$, $u(t)$ and $e(t)$. The output $y(t)$ is known and the signal $u(t)$ is at our disposal. The variable $e(t)$ can be computed from past inputs and outputs. Choosing the variable $u(t)$ so that the terms underlined in equation (4.1) vanishes thus makes the variance of $y(t+1)$ as small as possible. This gives (4.2) and (4.3). For further details, see Åström (1970).

Since the process (4.1) is characterized by three parameters a straightforward explicit self-tuner would require estimation of three parameters. Estimation of the parameter c is also a nonlinear problem. Notice, however, that the feedback law is characterized by one parameter only. A self-tuner which estimates this parameter can be obtained based on the model

$$y(t+1) = \theta y(t) + u(t) \tag{4.4}$$

The least squares estimate of the parameter θ in this model is given by

$$\theta(t) = \frac{\sum_{k=1}^{t} y(k)\,[y(k+1) - u(k)]}{\sum_{k=1}^{t} y^2(k)} \tag{4.5}$$

and the control law is then given by

$$u(t) = -\theta(t)y(t) \tag{4.6}$$

The self-tuning regulator given by (4.5) and (4.6) has some remarkable properties which can be seen heuristically as follows. Equation (4.5) can be written as

$$\frac{1}{t}\sum_{k=1}^{t} y(t+1)y(t) = \frac{1}{t}\sum_{k=1}^{t} [\theta(t)y^2(k) - u(k)y(k)] = \frac{1}{t}\sum_{k=1}^{t} [\theta(t) - \theta(k)]y^2(k)$$

Assuming that y is mean square bounded and that the estimate $\theta(t)$ converges as $t \to \infty$ we get

$$\lim \frac{1}{t}\sum_{k=1}^{t} y(k+1)y(k) = 0 \tag{4.7}$$

The adaptive algorithm (4.5), (4.6) thus attempts to adjust the parameter θ so that the correlation of the output at lag one is zero. If the system to be controlled is actually governed by (4.1) it follows from (4.3) that the estimate will converge to the minimum variance control law under the given assumption. This is somewhat surprising because the structure of (4.4) which was the basis of the adaptive regulator is not compatible with the true system (4.1). More details are given in Åström and Wittenmark (1973,1985) □

Indirect Adaptive Control

An advantage of indirect adaptive control is that many different design methods can be used. The key issue in analysis of the indirect schemes is to show that the parameter estimates converge. This will in general require that the model structure used is appropriate and that the input signal is persistently exciting. To ensure this it may be necessary to introduce perturbation signals. Provided that proper excitation is provided there are no difficulties in controlling an integrator whose gain may have different sign.

Direct Adaptive Control

The direct adaptive control schemes may work well even if the model structure used is not correct as wa shown in Example 3.1. The direct schemes will, however, require other assumptions. Assume e.g. that the process to be controlled can be described by

$$A(q)y(t) = B(q)u(t) + v(t) \tag{4.8}$$

where u is the input, y is the output, v is a disturbance and A9q) and B(q) are polynomials in the forward shift operator. Stability of adaptive control of (4.8) have been given by Egardt (1979), Fuchs (1979), Goodwin et al. (1980), Gawthrop (1980), de Larminat (1979), Morse (1980), and Narendra et al. (1980). So far the stability proofs are available only for some simple algorithms. The following assumptions are crucial:

(A1) the relative degree d = deg A - deg B is known,

(A2) the sign of the leading coefficient b_0 of the polynomial B(q) is known,

(A3) the estimated model is at least of the same order as the process,

(A4) the polynomial B has all zeros inside the unit disc.

The assumption A1 means that the time delay is known with a precision, which corresponds to a sampling period. This is not unreasonable. For continuous time systems the assumption means that the Together with assumption (A2) it also means that the phase is known at high frequencies. If this is the case, it is possible to design a robust high gain regulator for the problem, see Horowitz (1963), Horowitz and Sidi (1973), Leitmann (1979) and Gutman (1979). For many systems like flexible aircraft, electromechanical servos and flexible robots, the main difficulty in control is the uncertainty of the dynamics at high frequencies, see Stein (1980).

Assumption A3 is very restrictive, since it implies that the estimated model must be at least as complex as the true system, which may be nonlinear with distributed parameters. Almost all control systems are in fact designed based on strongly simplified models. High frequency dynamics are often neglected in the simplified models.

Assumption A4 is also crucial. It arises from the necessity to have a model, which is linear in the parameters in the direct schemes.

5. ROBUST ADAPTIVE CONTROL

For a long time the research on stability of adaptive control systems focussed on proofs of global stability for all values of the adaptation gain. The results obtained under such premises are naturally quite restrictive. To get some insight into this consider a continuous time system described by

$$y = G(p)u \tag{5.1}$$

where u is the input, y is the output, G is the transfer function of the system and p = d/dt is the differential operator. Consider also the model reference adaptive control law given by

$$\begin{aligned} u &= \theta^T \varphi \\ \frac{d\theta}{dt} &= -k\varphi e \\ e &= y - y_m \end{aligned} \tag{5.2}$$

where y_m is the desired model output, e the error and θ a vector of adjustable parameters. The components of the vector φ are functions of the command signal. In a simple case, where the regulator is a combination of a proportional feedback and a proportional feedforward, φ becomes

$$\varphi = [r \ -y]^T$$

where r is the reference signal.

It follows from (5.1) and (5.2) that

$$\frac{d\theta}{dt} + k\varphi[G(p)\varphi^T\theta] = k\varphi y_m \tag{5.3}$$

This equation gives insight into the behavior of the system. Assume that the adaptation loop is much slower than the process dynamics. The parameters then change much slower than the regression vector φ and the term $G(p)\varphi^T\theta$ in (3.3) can then be approximated by its average i.e.

$$G(p)\varphi^T\theta \approx [\overline{G(p)\varphi^T(\theta)}]\theta \tag{5.4}$$

where "———" denotes time averages. Notice that the regression vector φ depends on the parameters. The following approximation to (5.3) is obtained

$$\frac{d\theta}{dt} + k\overline{\varphi(\theta)[G(p)\varphi^T(\theta)]}\theta \approx k\varphi y_m \tag{5.5}$$

This is the normal situation because the adaptive algorithm is motivated by the fact the parameters change slower than the other variables in the system under this assumption. Notice, howerer, that it is not easy to guarantee that the parameters change slowly by choosing k sufficiently small.

Equation (5.4) is stable if $k\varphi[G(p)\varphi^T]$ is positive. This is true e.g. if G is strictly positive real and if the input signal is persistently exciting. However, if the transfer function G(s) is strictly positive real it is also possible to design a robust high gain feedback for the system. We thus arrive at the paradox that the assumption required to show stability of the adaptive system will allow the design of a robust feedback. The assumption that G(s) is strictly positive real is, however, not necessary as is shown by the following example.

Example 5.1

Consider a system where only a feedforward gain is adjusted and let the command signal be a sum of sinusoids i.e.

$$r(t) = \sum_{k=1}^{n} a_k \sin(\omega_k t)$$

Using the model reference algorithm given by (5.2) the parameter estimates satisfy

$$\frac{d\theta}{dt} = kr[1 - \theta]G(p)r$$

Assuming that the gain is small and using averages we find that the estimates are approximately given by

$$\frac{d\bar{\theta}}{dt} = ka[1 - \bar{\theta}] \tag{5.6}$$

where

$$a = \frac{1}{2}\sum_{k=1}^{n} a_k^2 \cos[\arg G(i\omega_k)] \tag{5.7}$$

The equation (5.6) is stable if a is positive. Consider first the case of a single sinusoidal, n = 1, the equation is then unstable if the frequency of the command signal is chosen so that $G(i\omega_n)$ has a phase-shift larger than 90°. If the input contains several frequencies it is necessary that the dominating contribution to (5.7) comes from frequencies where the phase of $G(i\omega)$ is less than 90°. □

6. UNIVERSAL STABILIZERS

Adaptive control systems are nonlinear systems with a special structure. They are often designed based on the idea of automating modeling and design. It is natural to ask if there are other types of nonlinear controls which also can deal with uncertainties in the process model. A special class of systems were generated as attempts of solving the following problem which was proposed by Morse (1983). Consider the system

$$\frac{dy}{dt} = ay + bu$$

where a and b are unknown constants. Find a feedback law of the form

$$u = f(\theta, y)$$

$$\frac{d\theta}{dt} = g(\theta, y)$$

which stabilizes the system for all a and b. Morse conjectured that there are no rational f and g which stabilize the system. Morse's conjecture was proven by Nussbaum (1983) who also showed that there exist nonrational f and g which stabilize the system, e.g. the following functions

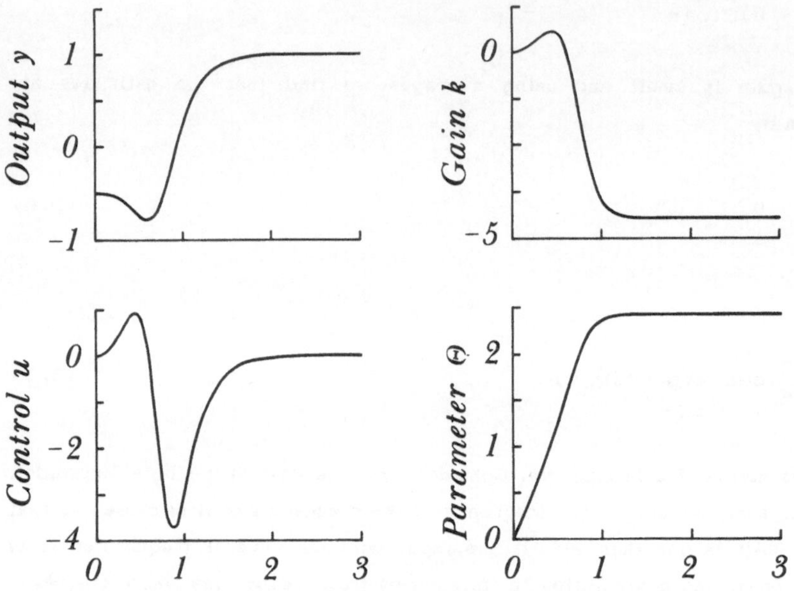

Figure 6. Simulation of an integrator with Nussbaum's control law.

$$f(\theta, y) = y\theta^2 \cos\theta$$

$$g(\theta, y) = y^2$$

This correspond to proportional feedback with the gain

$$k = \theta^2 \cos\theta$$

Figure 6 shows a simulation of this control law applied to an integrator with unknown gain. Notice that the regulator is initialized so that the gain has the wrong sign. In spite of this the regulator recovers and changes the gain appropriately. Nussbaum's regulator is of considerable principal interest because it shows that the assumption A2 is not necessary. The control law is, however, not necessarily a good control law in a practical situation because it may generate quite violent control actions. The initial conditions for the simulation shown in Figure 6 were chosen quite carefully.

7. DUAL CONTROL THEORY

Uncertainties can also be captured using nonlinear stochastic control theory. The system and its environment are then described by a stochastic model. To do so the parameters are introduced as state variables and the parameter uncertainty is modeled by stochastic models. An unknown constant is thus modeled by the differential equation

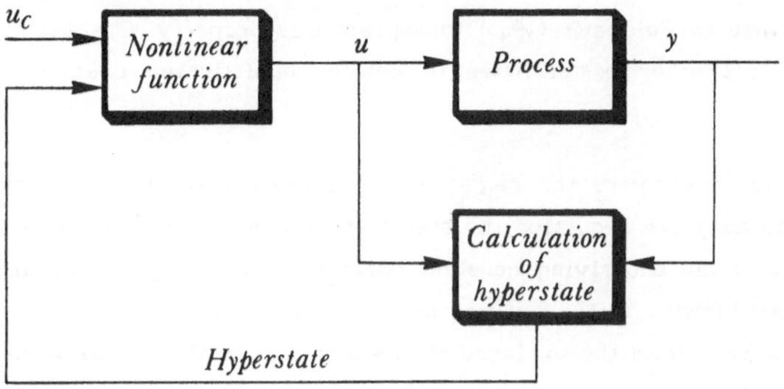

Figure 7. Block diagram of an adaptive regulator obtained from stochastic control theory.

$$\frac{d\theta}{dt} = 0$$

with an initial distribution that reflects the parameter uncertainty. Parameter drift is modeled by adding random variables to the right hand sides of the equations. A criterion is formulated as to minimize the expected value of a loss function, which is a scalar function of states and controls.

The problem of finding a control, which minimizes the expected loss function, is difficult. Under the assumption that a solution exists, a functional equation for the optimal loss function can be derived using dynamic programming, see Bellman (1957,1961). The functional equation, which is called the Bellman equation, is a generalization of the Hamilton-Jacobi equation in classical variational calculus. It can be solved numerically only in very simple cases. The structure of the optimal regulator obtained is shown in Figure 7. The controller can be thought of as composed of two parts: a nonlinear estimator and a feedback regulator. The estimator generates the conditional probability distribution of the state from the measurements. This distribution is called the hyperstate of the problem. The feedback regulator is a nonlinear function, which maps the hyperstate into the space of control variables. This function can be computed off-line. The hyperstate must, however, be updated on-line. The structural simplicity of the solution is obtained at the price of introducing the hyperstate, which is a quantity of very high dimension. Updating of the hyperstate requires in general solution of a complicated nonlinear filtering problem. Notice that there is no distinction between the parameters and the other state variables in Figure 7. This means that the regulator can handle very rapid parameter variations.

The optimal control law has interesting properties which have been found by solving a number of specific problems. The control attempts to drive the output to its desired value, but it will also introduce perturbations (probing) when the parameters are uncertain. This improves the quality of the estimates and the future controls. The optimal control gives the correct balance between maintaining good control and small estimation errors. The name dual

<u>control</u> was coined by Feldbaum (1965) to express this property. Optimal stochastic control theory also offers other possibilities to obtain sophisticated adaptive algorithms, see Saridis (1977).

It is interesting to compare the regulator in Figure 7 with the self-tuning regulator in Figure 5. In the adaptive regulator the states are separated into two groups, the ordinary state variables of the underlying constant parameter model and the parameters which are assumed to vary slowly. In the optimal stochastic regulator there is no such distinction. There is no feedback from the variance of the estimate in the adaptive regulator although this information is available in the estimator. In the optimal stochastic regulator there is feedback from the conditional distribution of parameters and states. The design calculations in the adaptive regulator is made in the same way as if the parameters were known exactly. Finally there are no attempts in the adaptive regulator to introduce the estimates when they are uncertain. In the optimal stochastic regulator the control law is calculated based on the hyperstate which takes full account of uncertainties. This also introduces perturbations when estimates are poor. The comparison indicates that it may be useful to add parameter uncertainties and probing to the adaptive regulator. A simple example illustrates the dual control law and some approximations.

<u>Example 7.1</u>

Consider a discrete time version of the integrator with unknown gain

$$y(t+1) = y(t) + bu(t) + e(t), \tag{7.1}$$

where u is the control, y the output and e normal $(0,\sigma_e)$ white noise. Let the criterion be to minimize the mean square deviation of the output y. This is a special case of the system in Example 3.1 with a = 1 and c = 0. When the parameters are known the optimal control law is given by (3.2) i.e.

$$u(t) = -\frac{y(t)}{b} \tag{7.2}$$

If the parameter b is assumed to be a random variable with a Gaussian prior distribution, the conditional distribution of b, given inputs and outputs up to time t, is Gaussian with mean $\hat{b}(t)$ and standard deviation $\sigma(t)$. The hyperstate is then characterized by the triple $(y(t),\hat{b}(t),\sigma(t))$. The equations for updating the hyperstate are the same as the ordinary Kalman filtering equations, see Åström (1970) and (1978).

Introduce the loss function

$$V_N = \sigma_e^{-2} \min_u E\left\{ \sum_{k=t+1}^{t+N} y^2(k) \;\middle|\; Y_t \right\} \tag{7.3}$$

where Y_t denotes the data available at time t i.e. {y(t), y(t-1),...}. By introducing the normalized variables

$$\eta = y/\sigma_e, \quad \beta = \hat{b}/\sigma, \quad \mu = -u\hat{b}/y \tag{7.4}$$

it can be shown that V_N depends on η and β only. The Bellman equation for the problem can be written as

$$V_T(\eta,\beta) = \min U_T(\eta,\beta,\mu) \tag{7.5}$$

where

$$V_0(\eta,\beta) = 0$$

and

$$U_T(\eta,\beta,\mu) = (\eta-\mu\eta)^2 + 1 + \left(\frac{\mu\eta}{\beta}\right)^2 + \int_{-\infty}^{\infty} V_{T-1}(y,b)\varphi(\varepsilon)d\varepsilon \tag{7.6}$$

where φ is the normal probability density and

$$y = \eta - \mu\eta + \varepsilon\sqrt{1 + \left(\frac{\mu\eta}{\beta}\right)^2}$$

$$b = \frac{\mu\eta\varepsilon}{\beta} + \beta\sqrt{1 + \left(\frac{\mu\eta}{\beta}\right)^2} - \frac{\mu\beta}{\eta}\varepsilon$$

see Åström (1978). When the minimization is performed the control law is obtained as

$$\mu_T(\eta,\beta) = \arg\min_u U_T(\eta,\beta,\mu) \tag{7.7}$$

The minimization can be done analytically for T = 1. We get

$$\mu_1(\eta,\beta) = \arg\min_u \left[(\eta-\mu\eta)^2 + 1 + \left(\frac{\mu\eta}{\beta}\right)^2\right] = \frac{\beta^2}{1+\beta^2}$$

Transforming back to the original variables we get

$$u(t) = -\frac{1}{\hat{b}(t)} \cdot \frac{\hat{b}^2(t)}{\hat{b}^2(t)+\sigma^2(t)}\, y(t) \tag{7.8}$$

This control law is called one-step control or myopic control because the loss function V_1 only looks one step ahead.

For T > 1 the optimization can no longer be made analytically. Instead we have to resort to numerical calculations. For large values of T the solution can be approximated by

$$\mu(\eta, \beta) = \frac{0.56\beta + \beta^2}{2.2 + 0.08\beta + \beta^2} + \frac{1.9\beta}{\eta(1.7 + \beta^4)} \qquad \eta > 0, \ \beta > 0.$$

The control law is an odd function in η and β, see Åström and Helmersson (1983).

Some approximations to the optimal control law will also be discussed. The certainty equivalence control

$$u(t) = - y(t)/\hat{b} \tag{7.9}$$

is obtained simply by taking the control law (2.24) for known parameters and substituting the parameters by their estimates. The self-tuning regulator can be interpreted as a certainty equivalence control. Using normalized variables the control law becomes

$$\mu = 1 \tag{7.9'}$$

The myopic control law (7.8) is another approximation. This is also called cautious control, because in comparison with the certainty equivalence control it hedges and uses lower gain when the estimates are uncertain. Notice that all control laws are the same for large β i.e if the estimate is accurate. The optimal control law is close to the cautious control for large control errors. For estimates with poor precision and moderate control errors the dual control gives larger control actions than the other control laws.

A simulation of the dual control law for an integrator with variable gain is shown in Figure 8. Notice that the gain varies by an order of magnitude in size and that it changes sign at T = 2000. In spite of this the regulator have little difficulty in controlling the process. Also notice that the regulator does probing well before the gain changes time and that it jumps between caution and probing when the gain passes through zero.

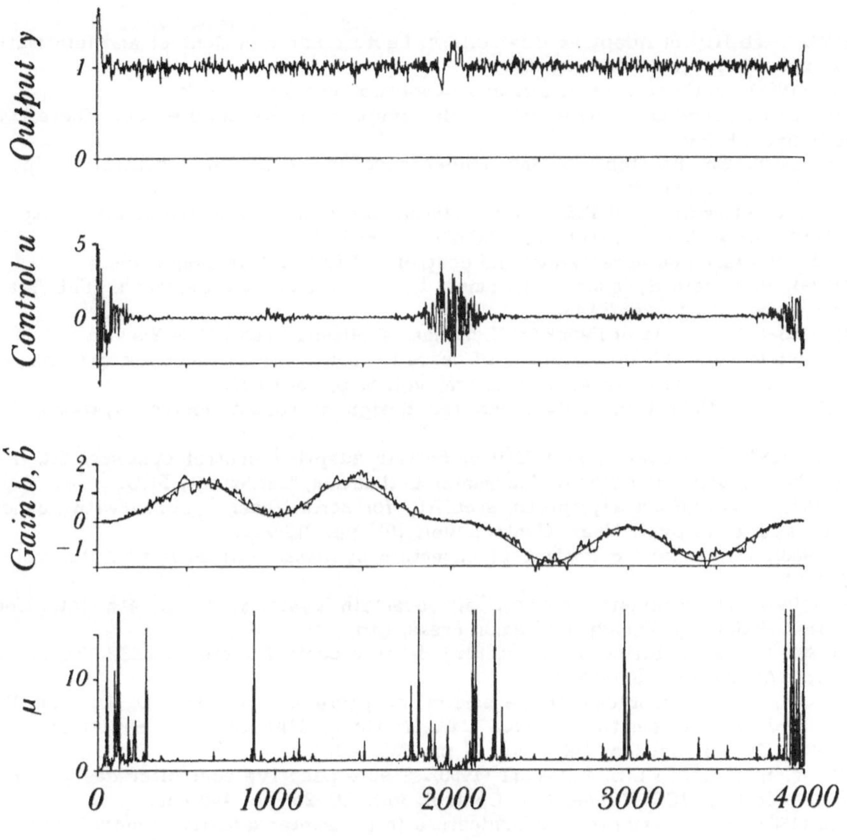

Figure 8. Simulation of dual control law applied to integrator with variable gain.

REFERENCES

Aström, K.J. (1970). Introduction to Stochastic Control Theory. Academic Press, New York.

Aström, K.J. (1978). Stochastic Control Problems. In Coppel, W.A. (Ed.). Mathematical Control Theory. Lecture Notes in Mathematics, Springer-Verlag, Berlin.

Aström, K.J. (1982). Ziegler-Nichols auto-tuners. Report CODEN: LUTFD2/TFRT-3167, Dept. of Automatic Control, Lund Institute of Technology, Lund, Sweden.

Aström, K.J. (1983). Theory and applications of adaptive control - A survey. Automatica, vol. 19, pp. 471-487, 1983.

Aström, K.J., and T. Hägglund (1984a). Automatic tuning of simple regulators. Proceedings IFAC 9th World Congress, Budapest, Hungary.

Aström, K.J., and T. Hägglund (1984b). Automatic tuning of simple regulators with specifications on phase and amplitude margins. Automatica, vol. 20, No. 5, Special Issue on Adaptive Control, pp. 645-651.

Aström, K.J., and B. Wittenmark (1973). On self-tuning regulators. Automatica, vol. 9, pp. 185-199.

Aström, K.J., and B. Wittenmark (1985). The self-tuning regulators revisited. Proc. 7th IFAC Symp. Identification and System Parameter Estimation, York, UK.

Bellman, R. (1957). Dynamic Programming. Princeton University Press.

Bellman, R. (1961). Adaptive Processes - A Guided Tour. Princeton University Press.

Doyle, J.C and G. Stein (1981). Multivariable feedback design: Concepts for a Classical/Modern Synthesis. IEEE Trans. Aut. Control, vol. AC-20, pp. 4-16.

Egardt, B. (1979). Stability of Adaptive Controllers. Lecture notes in Control and Information Sciences, Vol. 20, Springer-Verlag, Berlin.

Feldbaum, A.A. (1965). Optimal Control System. Academic Press, New York.

Fuchs, J.J. (1979). Commande adaptative directe des systemes linéaires discrets. These D.E., Univ. de Rennes, France.

Gawthrop, P.J. (1980). On the stability and convergence of a self-tuning controller. Int. J. Control, vol. 31, pp. 973-998.

Goodwin, G.C., P.J. Ramadge, and P.E. Caines (1980). Discrete-time multivariable adaptive control. IEEE Trans. Aut. Control, vol. AC-25, pp. 449-456.

Grübel, G. (1985). Research on uncertainty and control at DFVLR. This symposium.

Gutman, S. (1979). Uncertain dynamical systems - Lyapunov min-max approach. IEEE Trans. Aut. Control, vol. 24, pp. 437-443.

Horowitz, I.M. (1963). Synthesis of Feedback Systems. Academic Press, New York.

Horowitz, I.M., and M. Sidi (1973). Synthesis of cascaded multipleloop feedback systems with large plant parameter ignorance. Automatica, vol. 9, pp. 589-600.

Kwakernaak, H. (1985). Uncertain models and the design of robust control systems. This symposium.

Larminat, Ph. de (1979). On overall stability of certain adaptive control systems. 5th IFAC Symp. on Identification and System Parameter Estimation, Darmstadt, FRG.

Leitmann, G. (1979). Guaranteed asymptotic stability for some linear systems with bounded uncertainties. J. Dyn. Syst. Meas. Control, vol. 101, pp. 212-216.

Leitmann, G. (1980). Deterministic control of uncertain systems. Astronautica Acta, vol. 7, pp. 1457-1461.

Leitmann, G. (1983). Deterministic control of uncertain systems. Proc. 4th Int. Conf. Mathematical Modelling, Zurich. Pergamon Press, Ltd.

Morse, A.S. (1980). Global stability of parameter-adaptive control systems. IEEE Trans. Aut. Control, vol. AC-25, pp. 433-439.

Morse, A.S. (1983). Recent problems in parameter adaptive control. In I.D. Landau, Ed., Outils et Modeles Mathematiques pour l'Automatique, l'Analyse de Systemes et le Traitement du Signal. Editions NRS, vol. 3, pp. 733-740.

Narendra, K.S., Y.-H. Lin, and L.S. Valavani (1980). Stable adaptive controller design - Part II: Proof of Stability. IEEE Trans. Aut. Control, vol. AC-25, pp. 440-448.

Nussbaum, R.D. (1983). Some remarks on a conjecture in parameter adaptive control. Systems and Control Letters, vol. 3, pp. 243-246, 1983.

Saridis, G.N. (1977). Self-Organizing Control of Stochastic Systems. Marcel Dekker, New York.

Stein, G. (1980). Adaptive flight control: A pragmatic view. In K.S. Narendra and R.V. Monopoli, Eds., Applications of Adaptive Control. Academic Press, New York, pp. 291-312.

Ziegler, J.G., and N.B. Nichols (1943). Optimum settings for automatic controllers. Trans. ASME, vol. 65, pp. 433-444.

OPTIMALITY IN ADAPTIVE CONTROL SYSTEMS

Ya.Z.Tsypkin

Institute of Control Sciences

Moscow, U.S.S.R.

1. Introduction

A major line of research in today's control system is control of dynamic plants under uncertainty, or design of adaptive control systems. Surveys of theory of adaptive systems are available such as Aseltine et al (1958), Stormer (1959), Donaldson and Kishi (1965), Landau (1974, 1981), Wittenmark (1975), Unbehauen and Schmidt (1975) and Isermann (1982). Adaptive system design and theory have been discussed in numerous books (Weber (1971), Eveleigh (1967), Sragovich (1976, 1981), Saridis (1977), Egardt (1979), Landau (1979), Petrov and Rutkovskiy (1980), Derevitskiy and Fradkov (1981), Fomin et al (1981), Ioannou and Kokotovic (1983), and Goodwin and Sin (1984)) and collected papers edited by Hammond (1966), Unbehauen (1980), Harris and Billings (1981), Narendra and Monopoli (1980), and chapters in the books by Åström (1970), Tsypkin (1971), Isermann (1981), Åström and Wittenmark (1983), Ljung and Söderström (1983), and Fomin (1984). An extensive list of references on the field is provided in Asher et al (1976). Adaptive systems were discussed in special issues of journals Proceedings of the IEEE (1976), IEEE Transactions on Information Theory (1984). The number of articles on adaptive systems seems to grow exponentially.

The advent of microcomputers gave new potential to discrete - time adaptive control systems (DACS) which are the chief subject of most literature on adaptive systems including the above References. Initially DACSs were classified into self-adjusting systems (SAS) and model reference adaptive systems (MRAS) (see Aseltine et al (1958), Unbehauen and Schmidt (1975), and Egardt (1979)). In the former the plant was identified and the controller parameters were then found from plant parameter estimates,in the latter the controller parameters were adjusted to make the closed-loop system akin to the reference model. But availability or absence of a reference model could not be a solid criterion, for in some way a reference model did exist in every adaptive system. More recent research reported adaptive systems with explicit or implicit reference model (Landau (1979)), with explicit or implicit self-adjustment (Åström (1980), Petrov and Rutkovskiy (1980)), explicit minimization of the criterion (Ljung and Trulsson (1981), Trulsson (1983) , and Tsypkin (1971)). Such classifications did not, however, necessarily follow from the principle of system operation and so did not contribute to consistency of the theory of adaptive systems, it was not until relatively recently that attempts were made to introduce more sound classification of adaptive systems (Proceedings of the IEEE (1976), IEEE Transactions on Information Theory (1984), Åström (1980), Ljung and Trulsson (1981), Trulsson (1983), Gawthrop (1977), Narendra and Valavani (1979), Johnson (1980)). Most adaptation algorithms were chosen heuristically. This is , probably why adaptation algorithms are fairly similar in the welter of papers. They respond linearly to errors or misalignments and are versions of recurrent stochastic approximation or least squares method. Modifications introduced in these algorithms often simplified the implementation or helped analytical study. Arbitrary choice of algorithms may result in failures. For this reason it is necessary to design adaptation algorithms and optimal , in some sense, algorithms in a consistent way. This however has been not on the agenda of researchers. To design control systems operated by optimal adaptive algorithms, it is first necessary to sort out the variety of concepts, definitions, structures and results of the DACS theory and then to introduce new concepts which would help to evaluate the performance and to optimize adaptation algorithms. When such algorithms are available the difference between their performance and that of other algorithms, in particular, those reported in the literature can be assessed.

Three kinds of optimality are offered by optimal DACSs of the control system structure, of the adjustable models, and of the adaptation algorithms. These kinds will be discussed below. To make the presentation straight forward and avoid numerous insignificant details, the discussion will concentrate on adaptive control of linear single input-single-output systems essentially without delays. Formal proof of convergence of the algorithms on which an impressive body of literature is available will also be skipped. The attention will focus on design of optimal algorithms and DACS properties.

2. Some History

The original adaptive control systems were self-adjusting and identifying the plant (SAS). That was natural , for if the plant parameters were available, the controller parameters would be easy to determine. The identification yielded the plant parameters and thus eliminated uncertainty resulting from lack of knowledge of them. The idea of SAS was proposed by Kalman (1958). The plant was identified by a least squares algorithm in recurrent form even though the plant was not subjected to noises. Algorithms of this kind were used in the presence of noise (Young (1965), Chang and Rissanen (1968), Peterka (1972), Åström and Wittenmark (1973)) and the chosen controllers resulted in SAS with minimal variance. This approach was later extended to more complicated quadratic criteria which allowed for control signals. SAS with required distribution of zeros and poles in the transfer function were considered by Clarke and Gawthrop (1975), Åström and Wittenmark (1980), Wellstead et al (1980), Wellstead and Sanoff (1981) and Clarke (1982).SASs and the various possibilities offered by identification of dynamic systems were studied in fairly numerous papers, including comparatively recent (Åström et al (1977), Narendra (1980), Kushner and Kumar (1982), Kumar and Moore (1982), Fuchs (1982 a,b), Kumar (1984), Caines and Lafortune (1984), Goodwin and Payne (1978), Ljung and Soderström (1983), and Tsypkin (1984)).

First model reference adaptive systems (MRAS) were continuous. The major objective of design was to eliminate or reduce the difference between outputs of the plant and the model which receive the same setpoint signal by varying the controller parameters. Such systems were studied by the second Lyapunov method.(Parks (1966) and later Petrov et al (1980)). Discrete-time MRASs were the subjects of numerous papers since Monopoli (1974) and Landau (1974), and then Ionescu and Monopoli (1977), Narendra and Valavani (1978), Landau (1979), Narendra and Lin (1980), Goodwin et al (1980, 1981), Landau and Lozano (1981), Cristi and Monopoli (1982) and Landau (1982a) . MRASs were often employed in the absence of noise. Recognition of the noise made it possible to determine how SASs and MRASs were related and what they had in common (Ljung and Landau (1980), Egardt (1979, 1980) and Landau (1981a, 1981b)). These researches studied chiefly the convergence of various algorithms modifications that responded to errors and misalignments linearly. Methods of common differential equations were used by ljung (1977a,b), Eweda and Macchio (1984), Solo

(1979), Goodwin et al (1981), Landau (1982a,b), etc. Error filtering for generality was used in Landau (1981a,1981b).

The conventional way to obtain a DACS, or a control system for a dynamic plant with unknown parameters, is to choose the control system structure and, consequently, that of controller with unknown parameters, and to choose an algorithm for estimating the plant parameters in a SAS or adjusting the controller parameters in a MRAS, and then to study convergence of the algorithms.

Even in the very first studies of adaptive systems much attention was given to choice of the control system structure. The chosen structures were, as a rule, optimal in that the quadratic functional of the error or misalignment was minimal (Åström (1970), Åström and Wittenmark (1973), Clarke and Gawthrop (1975), Gawthrop (1977), Landau (1979), Iserman (1981), Allidina and Hughes (1983), Goodwin and Sin (1984) etc.).

Wishing to cover various kinds of controllers in SASs and MRASs. Egardt (1979) introduced the notion of a filtered error which makes the controller restructarable to an extent. This does not reduce optimality because in fact a simpler quadratic criterion is replaced by a more sophisticated quadratic optimality critetion. Some authors choose the controller by obtaining the desired distribution of zeros and poles in the transfer function of the closed-loop system (e.g. Åström and Wittenmark (1980) , Wellstead and Sanoff (1981)). These approaches were compared by Clarke (1982). Adaptation algorithms respond, as a rule, linearly to errors or misalignment using various gain matrices and are,in fact, recurrent versions of the method of stochastic approximation and its modification (Panuska (1968), Saridis (1974), Tsypkin (1971), Derevitskiy and Fradkov (1981), Fomin and Fradkov (1981), Goodwin et al (1980,1981), Fu (1982)), of the method of least squares

(and its modifications) (Åström and Wittenmark (1973,1983), Åström et al (1977), Clarke and Gawthrop (1975), Kel'mans and Poznyak (1981), Sin and Goodwin (1982), De Klyser and Van Cauweuberghe (1983)), and the method of instrumental variables (Young and Hastings-James (1981)). In all these algorithms the gain matrix depends on current observations and on time. Sometimes this matrix is regarded as constant (Egardt (1979)) or introduce the "forgetting" factor (exponential weighting)(Wieslander and Wittenmark (1971), Allidina and Hughes (1981), Landau (1981b)). Many versions of adaptation algorithms are given in books by Landau (1979) and Goodwin and Sin (1984).

Adaptation algorithms are chosen on most occasions fairly arbitrarily; the choice is possibly dictated by various heuristic considerations which the authors make secret. This is probably why there is not much difference between algorithms reported in the numerous papers. Amendments of these classical algorithms had to be introduced to simplify either their analytical studies or implementation . In some papers the adaptation algorithms are fairly effectively tackle adaptive control. In some the algorithms result in biased estimates of the plant or controller parameters. While some authors do take note of this fact others overlook it and arrive at erroneous formulations. This is why it is exceedingly important to develop a technique of designing adaptive systems but not heuristically. This technique would make it possible to generate, in a standardized way, adaptive algorithms that would meet the requirements. For this purpose the basic DACS types should be described.

3. Basic Types of Adaptive Control Systems

The control system is a feedback system of a plant and a controller, Fig.1. In the Figure $y(n)$ is the plant output, $u(n)$ is the control signal, and $r(n)$ is the setpoint. The setpoint $r(n)$ is in one-to-one correspondence with a reference or digital quantity $y_0(n)$ and usually the control system has to reduce the error

$$e_0(n) = y(n) - y_0(n) . \tag{3.1}$$

The control system performance is usually estimated as an error functional of $e_0(n)$. There are numerous ways to design control systems, or to determine the parameters and the structure of the controller with the automatic system would meet the requirements. These control methods are employed with complete data available on the plant.

The need in adaptive systems arises when the data on the plant is incomplete. Adaptive control systems have to eliminate gradually the uncertainty arising from incomplete knowledge of plant parameters. These systems employ various adaptation algorithms which, as a result of processing the available observation data on inputs, outputs setpoints or reference values change the controller parameters so as to make the entire control system meet the requairements. Adaptive control systems can be obtained in several ways. In one the point is identified, or estimates θ of its parameters are obtained by identification algorithms which minimize thefunctional of misalignment $\varepsilon(n) = y(n) - \hat{y}(n)$ where $\hat{y}(n)$ is approximation of the plant output. A dedicated computer obtains from the estimates θ the controller parameters α which ensure the best possible performance of the entire system. When the estimates go to the actual values the control system structure goes to the desired one. The structure of such an adaptive system contains an identifier and computer in the adaptation loop (Fig.2). Another way is to determine directly the necessary controller parameter values. This is done either by algorithms predicting the desired reference values and minimizing the functional of misalignment $\varepsilon_0(n) = y(n) - \hat{y}_0(n)$ where now $\hat{y}_0(n)$ is approximation of the desired reference or by optimization algorithms minimizing the functional of error $e_0(n) = y(n) - y_0(n)$ of equation (3.1). The structure of an adaptive system incorporating a predictor is shown in Fig.3 and one with an optimizer in Fig.4. The difference between them is in the observation data used in the models to be tuned and in the algorithms which change the parameters of the model or of the controller. Unlike adaptive systems with identifiers, those with predictors and optimizers do not incorporate a computer. All these adaptive systems can be classified into indirect , where the controller parameters are found from estimates of the plant parameters by the computer and direct where the parameters are measured directly. The former include adaptive systems with an identifier, Fig.2. The latter include adaptive systems with a predictor, Fig.3, and those with an optimizer, Fig.4. Adaptive systems are feasible where the controller parameters are determined on the knowledge of some intermidiate quantities which depend on the plant parameters by some computing device. This is possible in control of dynamic systems with delays. Such adaptive systems will be classified with indirect systems.

All adaptive systems reported thus far, both SASs and MRASs can be classified, depending on their structure, either as indirect or direct. As noted above, in most papers adaptive systems are classified into SASs and MRASs and the latter, into explicit and implicit. Since relatively recently both SASs and MRASs are classified into explicit and implicit and into direct and indirect (Narendra and Valavani (1979)). Sometimes these classifications were made identical. Adaptive systems can also be classified into those with self-adjustable models (in identifiers or predictors) and those without self-adjustable models (in optimizers). The first classification above seems to be preferable as it represents significant features of adaptive systems rather than formal attributes such as presence or absence of a self-adjustable or reference model in the system.

4. Statement of the Problem

Let us consider a dynamic plant, Fig.5, incorporated into a control system, Fig. 1, and described by a linear difference equation

$$Q(q)\, y(n) = q\, P_u(q)\, u(n) + P_\xi(q)\, \xi(n) , \tag{4.1}$$

where $y(n)$ is the output quantity, $u(n)$ is the control action, $\xi(n)$ is the disturbance or noise, and q is a delay operator, $q^m x(n) = x(n-m)$, $m = 1, 2, \dots$. In most papers on adaptive systems the delay operator is denoted as q^{-1}: $q^{-m} x(n) = x(n-m)$, but we preferred a simpler and handier notation, q . Noise is a sequence of independent identically distributed random quantities such that

$$E\{\xi(n)\} = 0 , \quad E\{\xi(n)\,\xi(n-m)\} = \begin{cases} 0 & \text{with } m \neq 0 \\ \sigma_\xi^2 & \text{with } m = 0 . \end{cases} \tag{4.2}$$

$Q(q)$, $P_u(q)$ and $P_\xi(q)$ are polynomials such that

$$\begin{aligned} Q(q) &= 1 + a_1^* q + \dots + a_N^* q^N , \\ P_u(q) &= b_0^* + b_1^* q + \dots + b_{N_1}^* q^{N_1} , \\ P_\xi(q) &= 1 + c_1^* q + \dots + c_{N_2}^* q^{N_2} . \end{aligned} \tag{4.3}$$

The polynomials $P_u(q)$ and $P_\xi(q)$ are assumed unstable in that all their zeros stay outside a unit circle, $|q| \leq 1$. The dynamic plant has to be controlled so as to make the behavior of the control system with a stationary random setpoint $r(n)$ approach the desired behavior of the reference system Fig.6, which is described by the equation

$$G^0(q)\, y_0(n) = q H^0(q)\, r(n) \tag{4.4}$$

where $G^0(q)$ and $H^0(q)$ are specified stable polynomials of q of powers N_3 and N_4, respectively. In other words, the process $y(n)$ should differ as little as possible, in some sense, from the desired process $y_0(n)$ with the setpoint $r(n)$. The measure of this difference will be specified by the system performance criterion, or the functional

$$J = E\left\{\left[\frac{C^0(q)}{D^0(q)}\, e_0(n)\right]^2\right\}. \tag{4.5}$$

where $e_0(n) = y(n) - y_0(n)$ is the error of equation (3.1) and $C^0(q)$ and $D^0(q)$ are specified stable polynomials of powers N_5 and N_6, respectively. With the plant parameters known, or with the polynomials (4.3) known, what we have is the problem of designing an optimal controller (4.4). With an unknown plant it is the problem of designing an adaptive system.

Design of an optimal adaptive system consists of solving three simpler problems.

Problem 1. Design of an optimal system structure, or determining the structure of the optimal controller.

Problem 2. Design of optimal structures of self-adjustable identifier and predictor models.

Problem 3. Design of optimal adaptive algorithms, or identification, prediction, and optimization algorithms.

The performance of tunable models is estimated in terms of a functional such as (4.5) where the error is replaced by the misalignment $\varepsilon(n)$ or $\varepsilon_0(n)$ of identification, prediction, and optimization.

A natural estimate of the performance for adaptive algorithms is the rate of their convergence. Optimal algorithms converge the fastest. Note that Problems 2 and 3 close interrelated. Furthermore, whereas Problem 1 with the plant parameters assumed known has been thoroughly treated in the literature and resolved, Problem 2 and 3 have not come into the focus of attention until very recently. But it is in these problems that the specifics of various kinds of adaptive systems are felt.

Consequently, design of optimal adaptive systems should result in three types of optimality, of the control system structure, of the adjustable identifier and predictor models, and of the identification, prediction, and optimization algorithms.

5. Designing an Optimal Controller Structure

The diagram associated with this design is shown in Fig.7. It is required to determine the structure of controller units for which the functional (4.5) is at its minimal. This problem has been resolved when the plant parameters are known (Åström (1970), Isermann (1981)). The main finding will be presented below in a more general form convenient for the purposes of the discussion.

Theorem of an optimal controller.

For the plant (1.1) where the control is nonminimal-phase the optimal control minimizing the general criterion (4.5) is described by the equation

$$R(q)\, u(n) = P_r(q)\, r(n) - P_y(q)\, y(n) \tag{5.1}$$

where

$$\begin{aligned} R(q) &= D^0(q)\, G^0(q)\, P_u(q), \\ P_2(q) &= C^0(q)\, H^0(q)\, P_\xi(q), \\ P_y(q) &= G^0(q)\, P(q) \end{aligned} \tag{5.2}$$

where the polynomials $Q(q)$, $P_\xi(q)$, and

$$P(q) = g_0^* + g_1^* q + \dots + g_{N_7}^* q^{N_7}, \quad N_7 = max\,[N_2 + N_5 - 1, N_6 + N - 1] \tag{5.3}$$

satisfy the polinómial equation

$$C^0(q)\, P_\xi(q) = D^0(q)\, Q(q) + q\, P(q). \tag{5.4}$$

For optimal control

$$\frac{C^0(q)}{D^0(q)}\, e_0(n) = \frac{C^0(q)}{D^0(q)} \big(y(n) - y_0(n)\big) = \xi(n) \tag{5.5}$$

and the minimal value of the functional (4.5) is

$$J^* = \min_u J = E\left\{\left[\frac{C^0(q)}{D^0(q)}\, \xi(n)\right]^2\right\} = \sigma_\xi^2 . \tag{5.6}$$

The proof follows immediately from computation of the product $C^0(q)\, P_\xi(q)\, e_0(n)$ which, by virtue of the plant equation (4.1) and polynomial equation (5.4) reduces to the form

$$\begin{aligned} C^0(q)\, P_\xi(q)\, e_0(n) &= q\, [\, D^0(q)\, P_u(q)\, u(n) + \\ &+ P(q)\, y(n) - C^0(q)\, P_\xi(q)\, y_0(n+1)] + D^0(q)\, P_\xi(q)\, \xi(n). \end{aligned} \tag{5.7}$$

Equating the expression in brackets to zero we have

$$D^0(q)\, P_u(q)\, u(n) = C^0(q)\, P_\xi(q)\, y_0(n+1) - P(q)\, y(n). \tag{5.8}$$

Multiplying both sides of this equation by $G^0(q)$ and noting that from equation (4.4) follows the equality

$$G^0(q)\, y_0(n+1) = H^0(q)\, r(n) \tag{5.9}$$

we have in the notation of equation (5.2) an equation (5.1) for the controller and expressions for $\frac{C^o(q)}{D^o(q)} e_0(n)$, (5.5), and J^*, (5.6).

Representing equation (5.1) with an allowance for (5.2) in the form

$$u(n) = \frac{C^o(q)\, H^o(q)\, P_\xi(q)}{D^o(q)\, G^o(q)\, P_u(q)}\, r(n) - \frac{P(q)}{D^o(q)\, P_u(q)}\, y(n) \tag{5.10}$$

it is easy to determine the structure of an optimal controller, Fig.8. Denoting

$$\bar{r}(n) = \frac{C^o(q)\, H^o(q)}{D^o(q)\, G^o(q)}\, r(n) \quad \text{and} \quad \bar{y}(n) = \frac{1}{D^o(q)}\, y(n) \tag{5.11}$$

where the dash denotes a filtered quantity the equation of an optimal controller can be represented in the form

$$P_u(q)\, u(n) = P_\xi(q)\, \bar{r}(n) - P(q)\, \bar{y}(n). \tag{5.12}$$

With $\hat{P}_u(q)$, $\hat{P}_\xi(q)$, and $\hat{P}(q)$ denoting the polynomials (4.3) and (5.3) where the actual optimal coefficients are replaced by their estimates which will be denoted as $b_m\ (m = 0, 1, \dots, N_1)$, $c_m\ (m = 1, 2, \dots, N_2)$, $g_m\ (m = 0, 1, \dots, N_3 = N + N_2 - 1)$. The equation of such a recurrent controller where the parameters are not optimal takes the form

$$\hat{P}_u(q)\, u(n) = \hat{P}_\xi(q)\, \bar{r}(n) - \hat{P}(q)\, \bar{y}(n). \tag{5.13}$$

This controller equation may be represented in an explicitly recurrent form

$$u(n) = \frac{1}{b_0}\left[\bar{r}(n) - (\hat{P}_u(q) - b_0)\, u(n) - \hat{P}(q)\, \bar{y}(n) - (1 - \hat{P}_\xi(q))\, \bar{r}(n)\right]. \tag{5.14}$$

Equation (5.14) defines the optimal controller structure.

Let us introduce notation of the controller parameter and observation vectors

$$\alpha = (b_0, b_1, \dots, b_{N_1}, g_0, g_1, \dots, g_{N_7}, c_1, \dots, c_{N_2})^T, \tag{5.15}$$

$$\begin{aligned} z_\nu(n;s) = (&\nu u(n), u(n-1), \dots, u(n-N_1), \bar{y}(n), \dots, \bar{y}(n-N_7), \\ &-s(n-1+\nu), \dots, -s(n-N_2+\nu))^T \end{aligned} \tag{5.16}$$

where ν is either 0 or 1.

With $\nu = 0$ and $s(n) = \bar{r}(n)$ the observation vector

$$\begin{aligned} z_0(n;\bar{r}) = (&0, u(n-1), \dots, u(n-N_1), \bar{y}(n), \bar{y}(n-1), \dots, \\ &\bar{y}(n-N_7), -\bar{r}(n-1), \dots, -\bar{r}(n-N_2))^T \end{aligned} \tag{5.17}$$

is influenced by the variables in the right-hand side of equation (5.14). Consequently, the controller equation can be represented in a compact form

$$u(n) = \frac{1}{b_0}\left[\bar{r}(n) - \alpha^T z_0(n;\bar{r})\right]. \tag{5.18}$$

The dimension of the vector α and so of $z_\nu(n;s)$ is $N_\alpha = N_1 + N_2 + N_7 + 2$. This form of equations is most convenient for realization in adaptive systems.

6. Designing an Optimal Adjustable Identifier and Predictor Models

Solution of identification and prediction problems, as noted in Sect.3, rely on approximation $\hat{y}(n)$ of the plant output $y(n)$ and approximation $\hat{y}_0(n)$ of the desired or reference quantity. These approximations which depend on the plant parameters and thus on the reference are obtained from dedicated adjustable models in the identifier and the predictor. Before optimal structures of adjustable models are designed - their performance criteria have to be established. For identification this criterion may be the functional of misalignment

$$\varepsilon(n) = y(n) - \hat{y}(n) \tag{6.1}$$

or, more specifically,

$$J_I(\theta) = E\left\{[\varepsilon(n)]^2\right\}. \tag{6.2}$$

Theorem of optimal adjustable identifier model.

For a minimal-phase dynamic plant the optimal adjuctable identifier model is determined implicitly by the equation

$$P_\xi(q)\,\hat{y}(n) = P_u(q)\,u(n-1) + P(q)\,y(n-1) \tag{6.3}$$

where the polynomial $P(q)$ satisfies the polynomial equation

$$P_\xi(q) = Q(q) + q\,P(q) \tag{6.4}$$

or, in an explicit form, the equation

$$\hat{y}(n) = P_u(q)\,u(n-1) - (Q(q)-1)\,y(n) + (P_\xi(q)-1)\,\varepsilon(n). \tag{6.5}$$

Furthermore

$$\varepsilon(n) = \xi(n) \quad \text{and} \quad J_I^* = \min E\left\{[\varepsilon(n)]^2\right\} = \sigma_\xi^2. \tag{6.6}$$

Truth of this Theorem follows from direct computation of the product $P_\xi(q)\,\varepsilon(n)$ by virtue of the plant equation (4.1) and polynomial equation (6.4) once the terms independent of $\xi(n)$ is made equal to zero. Equation (6.5) is obtained by eliminating the polynomial $P(q)$ from equations (6.3) an. (6.4).

Let us introduce the notation of filtered misalignment and estimate

$$\bar{\varepsilon}(n) = S^0(q)\,\varepsilon(n), \quad \bar{\hat{y}}(n) = S^0(q)\,\hat{y}(n). \tag{6.7}$$

Once the optimal parameter values are replaced by their estimate the equation of an optimal adjustable identifier becomes

$$\hat{\bar{y}}(n) = P_u(q)\, u(n-1) - (Q(q)-1)\, y(n-1). \tag{6.8}$$

In a particular case $S^0(q)=1$ an optimal identifier model (6.8) was obtained in a different way by Tsypkin (1981). Nevertheless, some authors (Landau (1976 , 1984), Johnson (1984)) obtain obviously non-optimal adjustable models of identifier such as static

$$\hat{y}(n) = \hat{P}_u(q)\, u(n-1) - (\hat{Q}(q)-1)\, y(n-1)$$

or dynamic

$$\hat{y}(n) = \hat{P}_u(q)\, u(n-1) - (\hat{Q}(q)-1)\, \hat{y}(n-1).$$

Denoting the plant parameter estimate vector as

$$\theta = (b_0, b_1, \dots, b_{N_1}, -a_1, \dots, -a_N, c_1, \dots, c_{N_2})^T \tag{6.9}$$

and since the observation vector (5.21) is

$$\begin{aligned} z_1(n-1; -\bar{\varepsilon}) = (&u(n-1), u(n-2), \dots, u(n-N_1), \\ &-y(n-1), \dots, -y(n-N), \bar{\varepsilon}(n-1), \dots, \bar{\varepsilon}(n-N_2))^T \end{aligned} \tag{6.10}$$

the equation (6.8) describing the adjustable identifier model can be represented in a compact form

$$\hat{\bar{y}}(n) = \hat{\bar{y}}(n, \theta) = (S^0(q)-1)\, y(n) + \theta^T z_1(n-1; -\bar{\varepsilon}). \tag{6.11}$$

The dimension of θ and thus of $x(n-1)$ is $N_\theta = N + N_1 + N_2 + 1$. Let us now determine the optimal adjustable predictor model.

Let us denote the predicted by observations reference (desired) quantity $y_0(n)$ as $\hat{y}_0(n)$ and consider a controller equation of the form (5.8) but assume now that the output quantities are not $y_0(n)$ and $y(n)$ but $u(n)$ and $y(n)$. Then the output quantity is instead of a control signal, a quantity equal to the reference, setpoint $y_0(n+1)$ and, with non-optimal parameter values, the estimate $\hat{y}_0(n+1)$ .Consequently , when n is replaced by $n-1$ it follows from equation (5.8) that

$$C^0(q)\hat{P}_\xi(q)\hat{y}_0(n) = \mathcal{D}^c(q)\hat{P}_u(q)\,u(n-1) + \hat{P}(q)\,y(n-1) \tag{6.12}$$

and this predictor model is optimal by a criterion analogous with (4.5)

$$J_p(\alpha) = E\left\{\left[\frac{C^0(q)}{\mathcal{D}^c(q)}\,\varepsilon_0(n)\right]^2\right\} \tag{6.13}$$

where now

$$\varepsilon_0(n) = y(n) - \hat{y}_0(n) \tag{6.14}$$

is the misalignment. For a simpler predictor performance criterion such as (6.2)

$$J_p(\alpha) = E\left\{[S^0(q)\,\varepsilon_0(n)]^2\right\} \tag{6.15}$$

one should assume that $\mathcal{D}^0(q) \equiv 1$ and $C^0(q) = S^0(q)$ in equation (6.12). Then we have

$$S^0(q)\,P_\xi(q)\,\hat{y}_0(n) = P_u(q)\,u(n-1) + P(q)\,y(n-1). \tag{6.16}$$

Consequently, the theorem on optimal adjustable predictor model is true. For a minimal phase plant such a model is described by the equations

$$S^0(q)\,P_\xi(q)\,\hat{y}_0(n) = P_u(q)\,u(n-1) + P(q)\,y(n-1) \tag{6.17}$$

where $P(q)$ is a polynomial for which equation (6.4) is true. Furthermore,

$$S^0(q)\,\varepsilon_0(n) = \xi(n) \quad \text{and} \quad J_p^* = \min E\{[S^0(q)\,\varepsilon_0(n)]^2\} = \sigma_\xi^2. \tag{6.18}$$

What is important is that equation (6.17) of an optimal predictor coincides in this case of no delay in the plant with equation (6.3) in implicit form of an optimal adjustable model.

With the filtered predictor output denoted as

$$\hat{\bar{y}}_0(n) = S^0(q)\,\hat{y}_0(n) \tag{6.19}$$

the equation of the adjustable predictor model becomes

$$\hat{\bar{y}}_0(n) = P_u(q)\,u(n-1) + P(q)\,y(n-1) - (P_\xi(q)-1)\,\hat{\bar{y}}_0(n)\,. \tag{6.20}$$

With the notation of the controller parameter estimate vector α of equation (5.15) and the observation vector $z_\nu(n;s)$ of the equation (5.16) while $\nu = 1$ and $s(n) = \hat{\bar{y}}_0(n)$ and with n replaced by $n-1$, equation (6.19) becomes in compact form

$$\hat{\bar{y}}_0(n) = \hat{\bar{y}}_0(n,\alpha) = \alpha^T z_1(n-1;\hat{\bar{y}}_0)\,. \tag{6.21}$$

The parameters of adjustable identifier and predictor models should vary so as to become equal to those of the plant or to optimal controller parameters. This is achieved by adaptation algorithms, which are either identification or prediction algorithms or optimization algorithms which require neither identification nor prediction.

7. Optimality Criteria for Adaptation Algorithms

The accuracy of estimates $\theta(n)$ of the plant parameter vector θ^* that are generalized by algorithms is described by the error covariance matrix (ECM)

$$V_n = E\left\{(\theta(n)-\theta^*)(\theta(n)-\theta^*)^T\right\}. \tag{7.1}$$

No matter how they are determined , the unbiased estimates cannot converge to θ^* at a rate exceeding some probabilistic maximum which is described by the Cramer-Rao inequality (Cramer (1946), Rao (1965)) such as

$$V_n \geqslant \frac{1}{n I(p_0)}\, A^{-1}(\theta^*, \sigma^2(p_0)) \tag{7.2}$$

where

$$I(p_0) = E\left\{\left(\frac{p_0'(\xi)}{p_0(\xi)}\right)^2\right\} \tag{7.3}$$

is Fisher information depending on the error distribution density $p_0(\xi)$ and $A(\theta^*, \sigma^2(p_0))$ is a normalized information matrix which depends generally on the plant parameter vector θ^* and the noise variance $\sigma^2(p_0)$. The normalized information matrix is assumed positive definite and symmetrical, The right-hand side of the Cramer-Rao equation defines the ECM lower bound and so the maximal feasible accuracy of the estimate. Denote $V(n) = nV_n$ and determine the asymptotic error covariance matrix (AECM) as a limit.

$$V = \lim_{n\to\infty} V(n) = \lim_{n\to\infty} nV_n . \tag{7.4}$$

The AECM describes the asymptotic rate of convergence of the estimates and as follows from equations (7.4) and (7.2)

$$V \geqslant \frac{1}{I(p_0)} A^{-1}(\theta^*, \sigma^2(p_0)). \tag{7.5}$$

In a similar way for the estimates $\alpha(n)$ of optimal controller parameters

$$V_n = E\left\{(\alpha(n)-\alpha^*)(\alpha(n)-\alpha^*)^T\right\} \tag{7.6}$$

and

$$V \geqslant \frac{1}{I(p_0)} A^{-1}(\alpha^*, \sigma^2(p_0)). \tag{7.7}$$

The right-hand sides of the inequalities (7.5) and (7.7) which define the highest possible rate of convergence of the estimates $\theta(n)$ and $\alpha(n)$ depend only on the statistical characteristics of noise such as the distribution density $p_0(\xi)$ and variance $\sigma^2(p_0)$ whereas the left-hand sides depend on the way in which the estimates were obtained, or on the algorithms, and may serve as performance criteria.

The minimum of V which is equal to the right-hand sides of the inequalities (7.5) and (7.7) is associated with argument-optimal algorithms. It is natural that the values of performance criteria of identification $J_I(\theta)$ in equation (6.2), prediction $J_p(\alpha)$ in equation (6.15), and the control system $J(\alpha)$ in equation (4.5) depend on the associated estimates $\theta(n)$ and $\alpha(n)$ and these values can be used in describing the algorithms performance. Let us introduce algorithms performance criteria which are asymptotic deviations (AD)

$$w = \lim_{n\to\infty} E\left\{ J_I(\theta(n) - J_I^* \right\}, \tag{7.8}$$

in an analogous way

$$w = \lim_{n\to\infty} E\left\{ J_p(\alpha(n)) - J_p^* \right\} \tag{7.9}$$

and

$$w = \lim_{n\to\infty} E\left\{ J(\alpha(n)) - J_0 \right\}. \tag{7.10}$$

AD is minimal in optimal criterial algorithms. The objective of argument optimization is to determine optimal parameter vectors θ^* or α^* with which the associated criteria are minimal, whereas the criterial optimization is to achieve the minimum of criteria; there might be a set of optimal parameter vectors θ or α on which these minima are achieved. This is why in criterial optimization there is generally no need to determine the actua plant parameters or optimal controller parameters, but the criteria are nevertheless minimized. Criterial optimization imposes less restrictive conditions. Thus it does not require that the matrix $A(\cdot, \sigma^2(\rho_0))$ be positive definite or that an inverse matrix $A^{-1}(\cdot, \sigma^2(\rho_0))$ exist. If $A^{-1}(\cdot, \sigma^2(\rho_0))$ does than criterial and argument optimization yield the same effect.

8. Design of Optimal Identification Algorithms

Let us represent the identification performance criterion (6.2) in the form

$$J_I(\theta) = E\{\bar{\varepsilon}^2(n,\theta)\} \tag{8.1}$$

where by virtue of equations (6.7) and (6.11)

$$\bar{\varepsilon}(n,\theta) = S^0(q)\,\varepsilon(n,\theta) = S^0(q)\,y(n) - \hat{\bar{y}}(n,\theta)\,. \tag{8.2}$$

Equating the gradient of $J_I(\theta)$ to zero we have

$$\nabla_\theta J_I(\theta) = -2E\{\bar{\varepsilon}(n,\theta)\,v(n)\} = 0 \tag{8.3}$$

where the vector

$$v(n) = \nabla_\theta \hat{\bar{y}}(n,\theta) = S^0(q)\,\nabla_\theta \hat{y}(n,\theta) \tag{8.4}$$

is the sensitivity function. Computing $v(n)$ by equations of the adjustable identifier model (6.6) and of the dynamic plant we have

$$P_\xi(q)\,v(n) = z_1(n-1;-\bar{\varepsilon}) \tag{8.5}$$

or, in explicit recurrent form

$$v(n) = (1-\hat{P}_\xi(q))\,v(n) + z_1(n-1;-\bar{\varepsilon})\,. \tag{8.6}$$

With $\theta = \theta^*$, by virtue of the property (6.6) and independence of $\xi(n)$ and $v(n)$ the optimality condition can be generalized by introducing a nonlinear transformation of misalignment

$$-E\{\varphi(\bar{\varepsilon}(n,\theta))\,v(n)\} = 0 \tag{8.7}$$

where $\varphi(\varepsilon) = -\varphi(-\varepsilon)$ is an odd function. In compliance with the ideology of the adaptive approach of Tsypkin (1971), the condition (8.7) generates recurrent identification algorithm

$$\theta(n) = \theta(n-1) + \Gamma(n)\,\varphi[\bar{\varepsilon}(n, \theta(n-1))]\,\vartheta(n), \qquad \Gamma(n) = \frac{1}{n} B \tag{8.8}$$

where $B > 0$ is a positive definite matrix. The AECM which characterizes the rate of convergence of the algorithm (8.8) is defined by the matrix equation (Tsypkin (1984))

$$[\tfrac{1}{2} I - B \nabla_\theta^2 J_I(\theta)]\, V + V\, [\tfrac{1}{2} I - B \nabla_\theta^2 J_I(\theta)]^T + B R_A B^T = 0 \tag{8.9}$$

where

$$\nabla_\theta^2 J_I(\theta) = E\{\varphi'[\xi]\}\, A(\theta^*, \sigma^2(p_0)), \tag{8.10}$$

$$R_A = E\{\varphi^2[\xi]\}\, A(\theta^*, \sigma^2(p_0)) \tag{8.11}$$

and

$$A(\theta^*, \sigma^2(p_0)) = E\{\vartheta(n)\, \vartheta^T(n)\}. \tag{8.12}$$

The optimal matrix $B = B_0$ with which the AECM is at its minimal looks like

$$B_0 = [E\{\varphi'[\xi]\}\, A^{-1}(\theta^*, \sigma^2(p_0))]^{-1}. \tag{8.13}$$

Consequently, the optimal gain matrix takes the form

$$\Gamma(n) = \Gamma_0(n) = \frac{1}{n E\{\varphi'[\xi]\}} \left[A^{-1}(\theta^*, \sigma^2(p_0))\right]^{-1}. \tag{8.14}$$

Substituting B_0 from equation (8.13) in (8.4) yields the minimal value

$$V(\varphi, p_0) = \rho(\varphi, p_0)\, A^{-1}(\theta^*, \sigma^2(p_0)) \tag{8.15}$$

where the scalar multiplier is

$$\rho(\varphi, p_0) = \frac{E\{\varphi^2[\xi]\}}{\left[E\{\varphi'[\xi]\}\right]^2}. \tag{8.16}$$

The AD to the algorithm (8.8) is equal, with an optimal gain matrix $\Gamma_0(n)$ of equation (8.14), to

$$w(\varphi, p_0) = \rho(\varphi, p_0)\, N_\theta \tag{8.17}$$

where

$$N_\theta = \text{rank}\, A(\theta^*, \sigma^2(p_0)) \tag{8.18}$$

which is equal to the dimension of the plant parameter vector θ^*.

Unlike AECM , AD represents only the rank of the matrix $A(\theta^*, \sigma^2(p_0))$ which is equal to the dimension of the plant parameter vector. The scalar multiplier $\rho(\varphi, p_0)$ is a function of the non-linear transformation φ . Let us choose this non-linear transformation so as to minimize $\rho(\varphi, p_0)$ and thus both the AECM (8.15) and AD (18.17). Such an optimal non-linear transformation

$$\varphi_0(\bar{\varepsilon}) = \arg\min_{\varphi} \rho(\varphi, p_0) \tag{8.19}$$

depends on the noise distribution density $p_0(\xi)$ and is (Tsypkin (1984))

$$\varphi_0(\bar{\varepsilon}) = \left.\frac{d \ln p_0(\xi)}{d\xi}\right|_{\xi=\bar{\varepsilon}} = -\left.\frac{p_0'(\xi)}{p_0(\xi)}\right|_{\xi=\bar{\varepsilon}} . \tag{8.20}$$

With $\varphi(\bar{\varepsilon}) = \varphi_0(\bar{\varepsilon})$

$$\rho(\varphi_0, p_0) = \rho_{min} = E\{\varphi_0^2(\xi)\}^{-1} = E\left\{\left(\frac{p_0'(\xi)}{p_0(\xi)}\right)^2\right\}^{-1} = I^{-1}(p_0). \tag{8.21}$$

Consequently, minimal AECM and AD are

$$V(\varphi_0, p_0) = V(p_0) = \frac{1}{I(p_0)} A^{-1}(\theta^*, \sigma^2(p_0)) \tag{8.22}$$

and

$$w(\varphi_0, p_0) = w(p_0) = \frac{1}{I(p_0)} N_\theta \tag{8.23}$$

respectively. But equation (8.22) coincides with the expression (7.5) for the lowerbound of the Kramer-Rao inequality. Consequently, the recurrent algorithm (8.8) with the choice of an optimal gain matrix $\Gamma_0(n)$ in equation (8.14) and a nonlinear tranformation $\varphi_0(\bar{\varepsilon})$ in equation (8.20) can, by virtue of equations (8.6) and (8.12) , be represented in the form

$$\begin{aligned} \theta(n) &= \theta(n-1) + \Gamma_0(n)\, \varphi_0(\bar{\varepsilon}(n, \theta(n-1)))\, v(n), \\ v(n) &= (1 - \hat{P}_\xi(q))\, v(n) + z_1(n-1; -\bar{\varepsilon}), \\ \Gamma_0(n) &= \frac{1}{n I(p_0)} E\{v(n) v^T(n)\}^{-1}. \end{aligned} \tag{8.24}$$

Convergence of these algorithms is the fastest possible, for they are optimal both in the argument and criterial sense.

These identification algorithms cannot be physically implemented directly because the gain matrix $\Gamma_0(n)$ in equation (8.14) depends on an unknown vector Θ^* whereas equation (8.24) cannot be solved unless the statistical characteristics $v(n)$ are known and expected value of the matrix $v(n)\, v^T(n)$ is computed for every n. To obtain implementable algorithms the inverse gain matrix

$$\Gamma_0^{-1}(n) = n I(p_0)\, E\{v(n) v^T(n)\} = n E\{\varphi_0'(\xi)\, v(n) v^T(n)\} \tag{8.25}$$

is replaced by its estimate

$$\hat{\Gamma}_0^{-1}(n) = \sum_{m=1}^{n} \varphi_0'(\bar{\varepsilon}(m))\, v(m)\, v^T(m)$$

where follows the recurrent relationship

$$\hat{\Gamma}_0^{-1}(n) = \hat{\Gamma}_0^{-1}(n-1) + \varphi_0'(n)\, v(n)\, v^T(n). \tag{8.26}$$

From the well-known formula of matrix inversion (Ljung and Söderström (1983)) and equation (8.26) we have

$$\hat{\Gamma}_0(n) = \hat{\Gamma}_0(n-1) - \frac{\hat{\Gamma}_0(n-1)\, v(n)\, v^T(n)\, \hat{\Gamma}_0(n-1)}{(\varphi_0'[\varepsilon(n)])^{-1} + v^T(n)\, \hat{\Gamma}_0(n-1)\, v(n)}. \tag{8.27}$$

Consequently, realizable optimal identification algorithms can be represented as

$$\begin{aligned}
\Theta(n) &= \Theta(n-1) + \hat{\Gamma}_0(n)\, \varphi_0\left[\bar{\varepsilon}(n, \Theta(n-1))\right] v(n), \\
v(n) &= (1 - \hat{P}_\xi(q))\, v(n) + z_1(n-1; -\bar{\varepsilon}), \\
\hat{\Gamma}_0^{-1}(n) &= \hat{\Gamma}_0^{-1}(n-1) + \varphi_0'\left[\bar{\varepsilon}(n, \Theta(n-1))\right] v(n)\, v^T(n).
\end{aligned} \tag{8.28}$$

For simplification of the notation, in the subsequent algorithms equations (8.26) the inverse for the gain matrix (8.25) will be used and the algorithms for $\Theta(n)$ will not include a projector on the parameters $C(n)$ which is needed to stabilize the polynomial $\hat{P}_{\xi}(q)$.

The initial values for the optimal algorithms (8.28) are $\theta(0)=\theta_1$, and $v(m)=0$, $m=0,-1,\dots,-N_2$. It would be useful to have $\hat{\Gamma}_0^{-1}(0)=0$ and in equation (8.27) $\hat{\Gamma}_0(0)=I\cdot\infty$ where I is an identity matrix. In further description of the algorithms these initial values will be implied. The multiplier $\varphi'[\bar{\varepsilon}(n)]$ in the expression for $\hat{\Gamma}_0(n)$ facilitates obtaining estimates of specified accuracy with finite n , or accelerates the start of asymptotic behavior of the algorithms. In other words, the multiplier accelerates the algorithms. Examples of optimal nonlinear transformations $\varphi_0(\bar{\varepsilon})$ and their derivations $\varphi_0'(\bar{\varepsilon})$ in identification algorithms are shown in Table 1.

For normal distribution density $p_0(\xi)=N(0,\sigma_0^2)$

$$\varphi_0(\bar{\varepsilon}) = \frac{\bar{\varepsilon}}{\sigma_0^2} \quad \text{and} \quad \varphi_0'(\bar{\varepsilon}) = \frac{1}{\sigma_0^2} \tag{8.29}$$

and from equation (8.28) follow algorithms which respond linearly to the plant misalignment

$$\begin{aligned} \theta(n) &= \theta(n-1) - \hat{H}_0(n)\,\bar{\varepsilon}(n,\theta(n-1))\,v(n), \\ v(n) &= (1-\hat{P}_{\xi}(q))\,v(n-1) + z_1(n-1;-\bar{\varepsilon}), \\ \hat{H}_0^{-1}(n) &= \hat{H}_0^{-1}(n-1) + v(n)\,v^T(n). \end{aligned} \tag{8.30}$$

Note that

$$\hat{H}_0(n) = \frac{1}{\sigma_0^2}\,\hat{\Gamma}_0(n). \tag{8.31}$$

For Laplace's distribution density $p_0(\xi)=L(0,s)$ the nonlinear transformation $\varphi_0(\xi)=\operatorname{sign}\bar{\varepsilon}$ is switching and what we have are switching algorithms for the misalignment. Linear algorithms embody modifications of the method of least squares

(MLS) and switching algorithms, the method of least modules (MLM) (Mudrov and Kushko, (1976)). The procedure for generation of optimal adaptation algorithms reduces to determining the optimality condition for an optimal linear transformation and determining the sensitivity vector-functions from model or system equations.

9. Design of Optimal Prediction and Optimization Algorithms

Representing the prediction performance criterion (6.15) as

$$J_p(\alpha) = E\{\bar{\varepsilon}_0^2(n,\theta)\} \tag{9.1}$$

where, according to equations (6.7), (6.14), and (6.19)

$$\bar{\varepsilon}_0(n,\theta) = S^0(q)\,\varepsilon_0(n,\theta) = S^0(q)\,y(n) - \hat{\bar{y}}_0(n)\,. \tag{9.2}$$

A generalized optimality condition similar to equation (8.7) is

$$-E\{\varphi_0(\bar{\varepsilon}_0(n,\theta)\,v(n)\} = 0$$

here $\varphi_0(\bar{\varepsilon}_0)$ is an optimal nonlinear transformation and the sensitivity vector function v which was computed by the equation of adjustable reference prediction model is now described by the equation

$$P_\xi(q)\,v(n) = z_1(n-1;\bar{y}_0) \tag{9.3}$$

or

$$v(n) = (1 - P_\xi(q))\,v(n) + z_1(n-1;\hat{\bar{y}}_0)\,.$$

The procedure described in the preceding Sect.8 leads to optimal prediction algorithms

$$\begin{aligned} \alpha(n) &= \alpha(n-1) + \hat{\Gamma}_0(n)\,\varphi_0(\bar{\varepsilon}(n,\theta(n-1)))\,v(n), \\ v(n) &= (1 - P_\xi(q))\,v(n) + z_1(n-1;\hat{\bar{y}}_0), \\ \hat{\Gamma}_0^{-1}(n) &= \hat{\Gamma}_0^{-1}(n-1) + \varphi_0'(\bar{\varepsilon}(n,\theta(n-1)))\,v(n)\,v^T(n)\,. \end{aligned} \tag{9.4}$$

For a normal distribution density $\mathcal{N}(0, \sigma^2)$ we have algorithms linearly responding to deviations from the reference

$$\begin{aligned} \alpha(n) &= \alpha(n-1) + \hat{H}_0(n)\, \bar{\varepsilon}_0(n, \alpha(n-1))\, v(n), \\ v(n) &= (1 - \hat{P}_\xi(q))\, v(n) + z_1(n-1; \hat{\bar{y}}_0), \\ \hat{H}_0^{-1}(n) &= \hat{H}_0^{-1}(n-1) + v(n)\, v^T(n). \end{aligned} \tag{9.5}$$

In contrast to identification and prediction algorithms, in generation of optimization algorithms adjustable identifier or predictor models are unnecessary. True, this makes computation of the sensitivity vector-function somewhat more complicated but the form of the final result is not changed.

The optimization performance criterion naturally coincides with the control system performance criterion (4.5) and is represented in the form

$$J_0(\alpha) = E\{\tilde{e}_0^2(n, \alpha)\} \tag{9.6}$$

where

$$\tilde{e}_0(n,\alpha) = \frac{C^0(q)}{D^0(q)}\, e_0(n, \alpha) = \frac{C^0(q)}{D^0(q)}\, (y(n) - y_0(n)). \tag{9.7}$$

The parameter vector α dictates now the plant output $y(n)$ while $y_0(n)$ is dictated by the control signal $r(n)$ of equation (4.4). The generalized optimality condition has the form

$$E\{\varphi_0(\tilde{e}_0(n, \theta))\, v(n)\} = 0. \tag{9.8}$$

The sensitivity vector-function $v(n)$ is computed by the equation set (1.1), (5.1) with polynomial equation (5.9) by assumed true for parameter estimatès. This somewhat cumbersome computation yields

$$\hat{P}_{\xi}(q)\, v(n) = -\, z_1(n-1;\bar{\tau}) \tag{9.9}$$

or

$$v(n) = (1-\hat{P}_{\xi}(q))\, v(n) - z_1(n-1;\bar{\tau})$$

and the above procedure leads to optimal optimization algorithms

$$\begin{aligned} \alpha(n) &= \alpha(n-1) - \hat{\Gamma}_0(n)\, \varphi_0(\tilde{e}_0(n,\theta(n-1)))\, v(n), \\ v(n) &= (1-P_{\xi}(q))\, v(n) - z_1(n-1;\bar{\tau}), \\ \hat{\Gamma}_0^{-1}(n) &= \hat{\Gamma}_0^{-1}(n-1) + \varphi_0'(\tilde{e}_0(n,\theta(n-1)))\, v(n)\, v^T(n). \end{aligned} \tag{9.10}$$

For normal distribution density $\mathcal{N}(0, \sigma^2)$ we hand algorithms, linearly responding to errors,

$$\begin{aligned} \alpha(n) &= \alpha(n-1) - \hat{H}_0(n)\, \tilde{e}_0(n,\theta(n-1))\, v(n), \\ v(n) &= (1-\hat{P}_{\xi}(q))\, v(n) - z_1(n-1;\bar{\tau}), \\ \hat{H}^{-1}(n) &= \hat{H}^{-1}(n-1) + v(n)\, v^T(n). \end{aligned} \tag{9.11}$$

For optimal prediction and optimization algorithms equation (8.21) and (8.22) are replaced by the equations

$$V(\varphi_0, p_0) = V(p_0) = \frac{1}{I(p_0)}\, A^{-1}(\alpha^*, \sigma^2(p_0)) \tag{9.12}$$

and

$$w(\varphi_0, p_0) = w(p_0) = \frac{1}{I(p_0)}\, \mathcal{N}_{\alpha} \tag{9.13}$$

where $\mathcal{N}_{\alpha} = \mathrm{rank}\, A(\alpha^*, \sigma^2(p_0))$ which is equal to the dimension of the controller parameter vector α^*. In Sects 8 and 9 the matrix $A(\alpha^*, \sigma^2(p_0))$ was assumed to be positive definite , or inverse matrix $A^{-1}(\alpha^*, \sigma^2(p_0))$ was assumed to exist. If this assumption is not true and the matrix rank $\tilde{\mathcal{N}}_{\theta} < \mathcal{N}_{\alpha}$ and/or $\tilde{\mathcal{N}}_{\theta} < \mathcal{N}_{0}$ then the equality (9.12) makes no sense and argument convergence of the algorithms does not occur while the equality (9.13) is replaced by

$$w(\varphi_0, P_0) = w(P_0) = \frac{1}{I(P_0)} \tilde{N}_\alpha$$

which is evidence of optimal criterial convergence. The identification, prediction and optimization algorithms do not change form but the recurrent relationships for gain matrices perform pseudo-inversion rather than inversion at every step.

10. Optimal Adaptive Control Systems

Once the system, the adjustable models and the associated algorithms are optimized, the structure of indirect and direct adaptive systems can be described in detail and equations of their basic units obtained.

Indirect adaptive system (fig. 9)

The equation of the adjustable identifier or predictor model for the plant (6.11) becomes

$$\hat{\bar{y}}(n, \theta(n-1)) = (S^0(q) - 1)\, y(n) + \theta^T(n-1)\, z_1(n-1; -\bar{\varepsilon}) \tag{10.1}$$

The identification algorithms (8.28) take the form

$$\begin{aligned} \theta(n) &= \theta(n-1) + \hat{\Gamma}_0(n)\, \varphi_0(S^0(q)\, y(n) - \hat{\bar{y}}(n, \theta(n-1)))\, v(n), \\ v(n) &= (1 - P_\xi(q))\, v(n) + z_1(n-1; -\bar{\varepsilon}), \\ \hat{\Gamma}_0^{-1}(n) &= \hat{\Gamma}_0^{-1}(n-1) + \varphi_0'(S^0(q)\, y(n) - \hat{\bar{y}}(n, \theta(n-1)))\, v(n)\, v^T(n). \end{aligned} \tag{10.2}$$

The controller and plant parameters are related by polynomial equation (5.4) in the form

$$\hat{P}(q) = q^{-1}\left[C^0(q)\, \hat{P}_\xi(q) - D^0(q)\, \hat{Q}(q)\right]. \tag{10.3}$$

Controller equation (5.18) takes now the form

$$u(n) = \frac{1}{b_0(n)}\left[\bar{r}(n) - \alpha^T(n)\, z_0(n; \bar{r})\right]. \tag{10.4}$$

Note that the dimension of the identification algorithm (10.2) depends on that of the plant parameter vector

$$N_\theta = N + N_1 + N_2 + 1 . \tag{10.5}$$

The polynomial equation determines the unknown polynomial $P(q)$ of power

$$N_7 = \max\left(N_2 + N_5 - 1 ,\ N_6 + N - 1\right). \tag{10.6}$$

The dimension of the controller parameter vector is

$$N_\alpha = N_1 + N_2 + N_7 + 2 . \tag{10.7}$$

In general, N_α of equation (10.7) unlike N_θ of equation (10.5), as it can be seen from equation (10.6), depends on choice of the control system performance criterion, or on N_5 and N_6 .

Direct adaptive system with a predictor (Fig. 10).

The equation of a reference predictor model (6.21) takes the form

$$\hat{\bar{y}}_0(n, \alpha(n-1)) = \alpha^T(n-1)\, z_1(n-1; \hat{\bar{y}}_0) . \tag{10.8}$$

The prediction algorithms (9.4) become

$$\begin{aligned} \alpha(n) &= \alpha(n-1) + \hat{\Gamma}_0(n)\varphi_0\left[S^0(q)\, y(n) - \hat{\bar{y}}_0(n, \theta(n-1))\right] v(n), \\ v(n) &= (1 - P_\xi(q))\, v(n) + z_1(n-1; \hat{\bar{y}}_0), \\ \hat{\Gamma}_0^{-1}(n) &= \hat{\Gamma}_0^{-1}(n-1) + \varphi_0'\left[S^0(q)\, y(n) - \hat{\bar{y}}_0(n, \theta(n-1))\right] v(n)\, v^T(n). \end{aligned} \tag{10.9}$$

The controller equation (5.21) now is

$$u(n) = \frac{1}{b_0(n)}\left[\bar{r}(n) - \alpha^T(n)\, z_0(n; \bar{r})\right] . \tag{10.10}$$

Direct system with an optimizer (Fig.11).

The adaptation algorithms (9.10) in this case take the form

$$\alpha(n) = \alpha(n-1) - \hat{\Gamma}_0(n)\,\varphi_0[S^0(q)(y(n) - y_0(n))]\,v(n),$$
$$\vartheta(n) = (1 - \hat{P}_\xi(q))\,v(n) - z_1(n-1;\bar{r}), \tag{10.11}$$
$$\hat{\Gamma}_0^{-1}(n) = \hat{\Gamma}_0^{-1}(n-1) + \varphi_0'[S^0(q)(y(n) - y_0(n))]\,v(n)v^T(n).$$

Here $y(n) = y(n, \alpha(n-1))$ by virtue of closed-loop control system equations.

The controller equation is

$$u(n) = \frac{1}{b_0(n)}\left[\bar{r}(n) - \alpha^T(n) z_0(n;\bar{r})\right]. \tag{10.12}$$

The complexity of an adaptive system may be described in terms of algorithms dimension in the case of indirect adaptive systems (10.5) as

$$N_\theta = N + N_1 + N_2 + 1$$

and for direct adaptive systems (10.7) as

$$N_\alpha = N_1 + N_2 + N_7 + 2\,.$$

But it follows from equation (10.6) that

$$N_7 \geqslant N_6 + N - 1\,. \tag{10.13}$$

Therefore

$$N_\alpha \geqslant N_\theta + N_6\,. \tag{10.14}$$

The equality

$$N_\alpha = N_\theta \tag{10.15}$$

is true when only

$$N_6 = 0 \quad \text{and} \quad N_2 + N_5 \leqslant N\,. \tag{10.16}$$

Thence it follows that the complexity of direct adaptive systems cannot be lower than that of indirect adaptive systems if the need to solve a polynomial equation in the latter case is neglected.

11. Optimal Algorithms with Incomplete Prior Data.

The distribution density $p_0(\xi)$ of the noise $\xi(n)$ has been assumed known above. Such data is often lacking and only incomplete data on the distribution density is available such as the class to which $p_0(\xi)$ belongs (Tsypkin (1984)). These classes are, for example,

1. The class of all distribution densities

$$\mathcal{P}_1 = \left\{ p_0(\xi) : \; p_0(0) = \frac{1}{2s_0} > 0 \right\} \tag{11.1}$$

2. The class of distribution densities with limited variance

$$\mathcal{P}_2 = \left\{ p_0(\xi) : \int_{-\infty}^{\infty} \xi^2 p_0(\xi)\, d\xi \leq \sigma_1^2 \right\} \tag{11.2}$$

3. The class of nearly-normal distribution densities

$$\mathcal{P}_3 = \left\{ p_0(\xi) : \; p_0(\xi) = (1-\lambda)\, p_N(\xi) + \lambda p_1(\xi) \right\} \tag{11.3}$$

where $p_N(\xi) = N(0, \sigma_0^2)$ is a normal distribution density, $p_1(\xi)$ is an arbitrary distribution density and $0 \leq \lambda \leq 1$

4. The class of nearly uniform distribution densities

$$\mathcal{P}_4 = \left\{ p_0(\xi) : \; p_0(\xi) = (1-\lambda)\, p_R(\xi) + \lambda p_1(\xi) \right\} \tag{11.4}$$

where $p_R(\xi) = R(0, \ell)$ is a uniform distribution density, $p_1(\xi)$ is an arbitrary distribution density and $0 \leq \lambda \leq 1$

5. The class of finite distribution densities.

$$\mathcal{P}_5 = \left\{ p_0(\xi) : \int_{-\ell}^{\ell} p_0(\xi)\, d\xi = 1 \right\} . \tag{11.5}$$

It is assumed therefore that $p_0(\xi) = p_0(-\xi)$ and $\int_{-\infty}^{\infty} p_0(\xi)\, d\xi = 1$. The number of these classes can be significantly increased by imposing additional constraints on the variance, ets.

Let us refer to the nonlinear transformation

$$\varphi_*(x) = - \left. \frac{p_*'(\xi)}{p_*(\xi)} \right|_{\xi = x} \tag{11.6}$$

as optimal on the class $\mathcal{P}$ if for any distribution density p_0 the inequality for traces of inverse AECMs

$$\operatorname{tr} V^{-1}(\varphi_*, p_0) \geqslant \operatorname{tr} V^{-1}(\varphi_*, p_*) \geqslant \operatorname{tr} V^{-1}(\varphi_0, p_*) \tag{11.7}$$

or for inverse ADs

$$w^{-1}(\varphi_*, p_0) \geqslant w^{-1}(\varphi_*, p_*) \geqslant w^{-1}(\varphi_0, p_*) \tag{11.8}$$

are true.

Equation (11.7) expresses the argument and equation (11.8) criterial principle of optimality on a class. The least favorable distribution densities $p_*(\xi)$ which satisfy the argument and criterial optimality principles are, respectively,

$$p_*(\xi) = \arg\min_{p_0 \in \mathcal{P}} \operatorname{tr} V^{-1}(\varphi_0, p_0), \tag{11.9}$$

$$p_*(\xi) = \arg\min_{p_0 \in \mathcal{P}} w^{-1}(\varphi_0, p_0). \tag{11.10}$$

We have from equation (11.6) $\varphi_0(\xi) = - \frac{p_0'(\xi)}{p_0(\xi)}$. Then it follows from equations (8.21) and (8.22) that

$$p_*(\xi) = \arg\min_{p_0 \in \mathcal{P}} I(p_0) \operatorname{tr} A(\cdot, \sigma^2(p_0)) \tag{11.11}$$

and

$$p_*(\xi) = \arg\min_{p_0 \in \mathcal{P}} I(p_0). \tag{11.12}$$

The argument optimization problem for the class (11.1) does not have accurate analytical solutions ; an exception is made by the class $\mathcal{P}_2$ for which is the normal distribution density $p_*(\xi) = \mathcal{N}(0, \sigma^2)$ (see Table 2). Examples of optimal nonlinear transformations which are criterially optimal on a class are summarized in Table 3. Adaptation algorithms, optimal on a class, can be obtained by replacing in the above, optimal algorithms

$\varphi_0(\cdot)$ by $\varphi_*(\cdot)$ and $\varphi_0'(\cdot)$ by $\varphi_*'(\cdot)$.

What is important is that adaptive algorithms linearly responding to misalignments and errors are optimal on a class for any noise of limited variance as well as optimal for normal noise.

12. Modified Algorithms

Let us have a look at the optimality condition (8.3) for the identification problem. Substituting $v(n)$ from equation (8.6) in equation (8.3) we have

$$\begin{aligned} \nabla_\theta J(\theta) = &-2\,\mathrm{E}\{\varphi_0(\bar{\varepsilon}(n,\theta))(1-\hat{P}_\xi(q))\,v(n)\} - \\ &-2\,\mathrm{E}\{\varphi_0(\bar{\varepsilon}(n,\theta))\; z_1(n-1;-\bar{\varepsilon})\} = 0 . \end{aligned} \tag{12.1}$$

With $\theta=\theta^*$ it follows from equation (6.7) that $\bar{\varepsilon}(n,\theta) = \mathcal{S}^0(q)\,\varepsilon(n,\theta) = \xi(n)$ while $\xi(n)$ and $(1-P(q))\,v(n)$ along with $\xi(n)$ and $z_1(n-1;-\bar{\varepsilon})$ are independent. This is why every addend in equation (12.1) vanishes with $\theta=\theta^*$. In this way modified algorithms generated by the optimality condition $\mathrm{E}\{\varphi_0(\bar{\varepsilon}(n,\theta))\; z_1(n-1;-\bar{\varepsilon})\} = 0$ can be obtained which do not incorporate a sensitivity function equation.

The optimality condition can also be represented as

$$\mathrm{E}\{\varphi_0(\bar{\varepsilon}(n,\theta))(1-\hat{P}_\xi(\lambda q)\,v(n)\} + \mathrm{E}\{\varphi_0(\bar{\varepsilon}(n,\theta))\; z_1(n-1;-\bar{\varepsilon})\} = 0 \tag{12.2}$$

where $0 \le \lambda \le 1$. The factor λ maps,as it were, estimates of the polynomial $\hat{P}_\xi(q)$ coefficients on the stability region. These conditions generate modified identification algorithms such as

$$\begin{aligned} \theta(n) &= \theta(n-1) + \hat{\Gamma}_0(n)\,\varphi_0[\bar{\varepsilon}(n,\theta(n-1))]\,v(n), \\ v(n) &= (1-\hat{P}_\xi(\lambda q))\,v(n) + z_1(n-1;-\bar{\varepsilon}), \\ \hat{\Gamma}_0^{-1}(n) &= \hat{\Gamma}_0^{-1}(n-1) + \varphi_0'[\bar{\varepsilon}(n,\theta(n-1))]\,v(n)\,v^T(n). \end{aligned} \tag{12.3}$$

In a similar way modified prediction algorithms are obtained

$$\begin{aligned} \alpha(n) &= \alpha(n-1) + \hat{\Gamma}_0(n)\,\varphi_0[\bar{\varepsilon}(n,\theta(n-1))]\,v(n), \\ v(n) &= (1-\hat{P}_\xi(\lambda q))\,v(n) + z_1(n-1;\hat{\bar{y}}_0), \\ \hat{\Gamma}_0^{-1}(n) &= \hat{\Gamma}_0^{-1}(n-1) + \varphi_0'[\bar{\varepsilon}(n,\theta(n-1))]\,v(n)\,v^T(n) \end{aligned} \tag{12.4}$$

as are modified optimization algorithms

$$\begin{aligned} \alpha(n) &= \alpha(n-1) + \hat{\Gamma}_0(n)\,\varphi_0[\bar{e}_0(n,\theta(n-1))]\,v(n), \\ v(n) &= (1-\hat{P}_\xi(\lambda q))\,v(n) - z_1(n-1;\bar{r}), \\ \hat{\Gamma}_0^{-1}(n) &= \hat{\Gamma}_0^{-1}(n-1) + \varphi_0'[\bar{e}_0(n,\theta(n-1))]\,v(n)\,v^T(n). \end{aligned} \tag{12.5}$$

Modified identification algorithms (12.3), prediction algorithms (12.4), and optimization algorithms (12.5) with $\lambda = 1$ become optimal identification algorithms (10.2), prediction algorithms (10.9), and optimization algorithms (10.11). Choice of λ nearly equal to 1 stabilizes the polynomial $\hat{P}_\xi(\lambda q)$. If in modified algorithms (12.3) - (12.5) $\lambda = 0$, then we have simplified identification algorithms

$$\begin{aligned} \theta(n) &= \theta(n-1) + \hat{\Gamma}_0(n)\,\varphi_0[\bar{\varepsilon}(n,\theta(n-1))]\,z_1(n-1;-\bar{\varepsilon}), \\ \hat{\Gamma}_0^{-1}(n) &= \hat{\Gamma}_0^{-1}(n-1) + \varphi_0'[\bar{\varepsilon}(n,\theta(n-1))]\,z_1(n-1;-\bar{\varepsilon})\,z_1^T(n-1;-\bar{\varepsilon}), \end{aligned} \tag{12.6}$$

prediction algorithms

$$\alpha(n) = \alpha(n-1) + \hat{\Gamma}_0(n)\,\varphi_0[\bar{\varepsilon}_0(n,\theta(n-1))]\,z_1(n-1;\hat{\bar{y}}_0),$$
$$\Gamma_0^{-1}(n) = \Gamma_0^{-1}(n-1) + \varphi_0'[\bar{\varepsilon}_0(n,\theta(n-1))]\,z_1(n-1;\hat{\bar{y}}_0)\,z_1^T(n-1;\hat{\bar{y}}_0), \quad (12.7)$$

optimization algorithms

$$\alpha(n) = \alpha(n-1) - \hat{\Gamma}_0(n)\varphi_0[\bar{e}(n,\theta(n-1))]\,z_1(n-1;\bar{y}_0),$$
$$\hat{\Gamma}_0^{-1}(n) = \hat{\Gamma}_0^{-1}(n-1) + \varphi_0'[\bar{e}(n,\theta(n-1))]\,z_1(n-1;\bar{y}_0)\,z_1^T(n-1;\bar{y}_0) \quad (12.8)$$

where, as before

$$\varepsilon(n,\theta) = S^0(q)\,y(n) - (S^0(q)-1)\,y(n) - \theta^T(n-1)\,z_1(n-1;-\bar{\varepsilon}),$$
$$\varepsilon_0(n,\theta) = S^0(q)\,y(n) - \alpha^T(n-1)\,z_1(n-1;\hat{\bar{y}}), \quad (12.9)$$
$$\bar{e}_0(n,\theta) = S^0(q)\,y(n) - \bar{y}_0(n).$$

The rate of convergence of these simplified algorithms is somewhat below that of associated optimal algorithms. Replacement in equations (12.3) - (12.8) of $\varphi_0(\cdot)$ by $\varphi_*(\cdot)$ and $\varphi_0'(\cdot)$ by $\varphi_*'(\cdot)$ leads to modified and simplified algorithms, "optimal on a class". For the class of limited variance what we have then are simplified

identification algorithms

$$\theta(n) = \theta(n-1) + \hat{H}_0(n)\,\bar{\varepsilon}(n,\theta(n-1))\,z_1(n-1;-\bar{\varepsilon}),$$
$$\hat{H}_0^{-1}(n) = \hat{H}_0^{-1}(n-1) + z_1(n-1;-\bar{\varepsilon})\,z_1^T(n-1;-\bar{\varepsilon}), \quad (12.10)$$

prediction algorithms

$$\alpha(n) = \alpha(n-1) + \hat{H}_0(n)\,\bar{\varepsilon}(n,\theta(n-1))\,z_1(n-1;\hat{\bar{y}}_0),$$
$$\hat{H}_0^{-1}(n) = \hat{H}_0^{-1}(n-1) + z_1(n-1;\hat{\bar{y}}_0)\,z_1^T(n-1;\hat{\bar{y}}_0) \quad (12.11)$$

and optimization algorithms

$$\alpha(n) = \alpha(n-1) - \hat{H}_0(n)\,\bar{e}_0(n,\theta(n-1))\,z_1(n-1;\bar{z}),$$
$$\hat{H}_0^{-1}(n) = \hat{H}_0^{-1}(n-1) + z_1(n-1;\bar{z})\,z_1^T(n-1;\bar{z}). \quad (12.12)$$

All identification, prediction , and control algorithms available by now are particular cases of these linear algorithms with different choices of $S^0(q)$ (most frequently $S^0(q) \equiv 1$) and of gain matrices and sometimes of adjustable models. The most frequently used recurrent algorithms of the method of least squares are simplified "optimal algorithms on the class $\mathcal{P}_2$ " and with $\rho_\xi(q) = 1$, simplified "optimal" algorithms.

Above dynamic plants without delay were considered the findings about which are presumed extendible to dynamic plants with delays (see for instance Landau (1979, 1981a)). This may hold in arbitrary choice of algorithms but not with optimal algorithms in a general case. For plants with delay only indirect DACSs can be made optimal. As for direct DACSs, this case has to be extended to the case of correlated noise.

Conclusions

Discrete-time adaptive control systems are classified into indirect and direct. In the former the plant is identified and the resultant estimates are employed for proper manipulation of the controller parameters. In the latter the controller parameters are varied directly on the knowledge of the observed plant state compared with the current reference prediction or with the difference between the plant output and the reference.

Various kinds of DACSs are studied and surveyed in numerous articles concerned with various adaptation algorithms.

Adaptive algorithms were compared chiefly by computer simulation because no theoretical comparison methods were available.

In this context it was necessary to develop a general DACS theory so that with specified data on the plant and exogenous signals and requirements to the control system the DACS structure could be unambiguously determined and adaptive algorithms generated.

This kind of theory became possible by extensive use of the notion of optimality of the structure of the controller, adjustable identifier and predictor models, and finally, of adaptive algorithms. Consequently, optimal DACSs must incorporate three kinds of optimality. That of the system structure and adjustable models is obtained by minimizing quadratic criteria such as loss functions. Adaptation algorithms are optimized by minimizing their (argumental or criterial) rate of convergence. Optimal DACSs make use of the entire available data on the plant and noise.

This approach eliminates uncertainty in the choice of identification, prediction , and optimization algorithms in DACSs and provides a univalent way to generate these algorithms and to feature their basic properties from the available prior data. This approach has now to be extended to more complicated plants and DACS performance criteria.

References

Allidina A.Y., Hughes F.M. (1983), A general adaptive scheme. International Journal of Control, vol.38, pp.1115-1120.

Aseltine J.A., Manchini A.R., Sarture C.W. (1958),A survey of adaptive control systems. IRE Transactions on Automatic Control, vol.3, pp.102-108.

Asher R.B., Andrisani D.II, Dorato P. (1976), Bibliography on adaptive control systems. Proceedings of the IEEE, vol. 64, pp.1226-1240.

Åström K.J. (1970), Introduction to Stochastic Control Theory. Academic Press, New York.

Åström K.J. (1980), Desicion principles for self-tuning regulators. In Unbehauen H. (Ed.)"Methods and Applications in Adaptive Control", Proceedings of an International Symposium, Bochum, 1980, Springer Verlag, Berlin,pp.1-20.

Åström K.J., Borisson U., Ljung L., Wittenmark B. (1977), Theory and application of self-tuning regulators. Automatica, vol.13, pp.457-476.

Åström K.J., Wittenmark B. (1973), On self-tuning regulator. Automatica, vol.7, pp. 195-199.

Åström K.J., Wittenmark B. (1980), Self-adjusting controller based on pole-zero plassement. IEE Proceedings, vol. 127, pt.D, pp.120-130.

Åström K.J., Wittenmark B. (1983), Computer controlled systems : theory and design. Prentice Hall Inc., Englewood Cliffs, New Jersey.

Caines P.E., Lafortune S.(1984), Adaptive control with recursive identification for stochastic linear systems. IEEE Transactions on Automatic Control, vol.AC-29,pp.312-321.

Chang A., Rissanen J. (1968), Regulation of incomplitely identified linear systems. SIAM Journal of Control, vol.6, pp.327-348.

Clarke D.W. (1982), Model following and pole-placement self-tuners in Optimal Control Applications and Methods, vol.3, pp.323-335.

Clarke D.W., Gawthrop P.J.(1975), Self-tuning controller. Proceedings of the IEE, vol. 122, pp.929-934.

Cramer H. (1946), Mathematical methods of statistics. Princeton University Press, Princeton.

Cristi R., Monopoli R.V.(1982), Computational aspects of discrete time adaptive control. IEEE Transactions on Automatic Control, vol. 27, pp.722-725.

Derevitskiy D.P., Fradkov A.L. (1981), Applied theory of discrete-time adaptive control systems. Nauka, Moscow (in Russian).

Donaldson D.P., Kishi J. (1965), Review of adaptive control systems. Theories and Techniques in Modern Control Systems Theory. Ed. C.T.Leondes, N.Y.

Egardt B.(1979), Stability of Adaptive Controllers. Springer Verlag, Berlin.

Egardt B. (1980), Unification of some discrete-time adaptive control scheme. IEEE Transactions on Automatic Control, vol. 25, pp.693-697.

Eveleigh W. (1967), Adaptive Control and Optimization Technik. Mc Graw Hill, New York.

Eweda E., Macchi O. (1984), Convergence of an adaptive linear estimation algorithm. IEEE Transactions on Automatic Control, vol.AC-29, pp.119-127.

Fomin V.N. (1984), Recurrent estimation and adaptive filtering . Nauka, Moscow (in Russian).

Fomin V.N., Fradkov A.L., Yakubovich V.A. (1981), Adaptive control of dynamic plants. Nauka, Moscow (in Russian).

Fu Ch.H.(1982), Self-tuning controller and its convergence under correlated noise. International Journal of Control, vol.35, pp.1051-1059.

Fuchs J.J. (1982a), Indirect stochastic adaptive control: the general delay-white noise case. IEEE Transactions on Automatic Control, vol.27, pp.219-223.

Fuchs J.J. (1982b), Indirect stochastic adaptive control: the general delay-colored noise case. IEEE Transactions on Automatic Control, vol.27, pp.470-472.

Gawthrop P.J. (1977), Some interpretation of a self-tuning controller. Proceedings of the IEE , vol. 124, pp.889-894.

Goodwin G.C., Payne R.L. (1978), Dynamic systems identification : experiment design and data analysis. Academic Press, New York.

Goodwin G.C., Ramadge P.J., Caines P.E. (1980), Discrete-time multivariable adaptive control. IEEE Transactions on Automatic Control, vol. 25, pp.449-456.

Goodwin G.C., Ramadge P.J., Caines P.E. (1981), Discrete-time stochastic adaptive control. SIAM Journal of Control and Optimization, vol.19, pp.829-853.

Goodwin G.C., Sin K.S. (1984), Adaptive Filtering, Prediction and Control, Prentice Hall, Inc. , Englewood Cliffs, New Jercey.

Goodwin G.C., Sin K.S., Salnya K.K. (1980), Stochastic adaptive control and prediction. The geheral delay-colored noise case. IEEE Transactions on Automatic Control, vol. 25, No.5, pp.946-980.

Hammond P.H. (Editor) (1966), Theory of Self-adapting Control System. Plenum Press, New York.

Harris C.J., Billings S.A. (Editors) (1981), Self-tuning and adaptive control. Theory and Application , Peter Peregrinus Ltd, London.

IEEE Transactions on Information Theory, 1984, vol.IT-30, No.3, Special Issue on Adaptive Filtering.

Ioannou P.A., Kokotovoc P.V. (1983) , Adaptive Systems with Reduced Models. Springer Verlag, Berlin.

Ionescu T., Monopoli R.W. (1977), Discrete model reference adaptive control with augmented error signal. Automatica, vol. 13, pp. 507-517.

Isermann R. (1981), Digital Control Systems. Springer Verlag, Berlin.

Isermann R. (1982), Parameter adaptive control algorithms.A tutorial. Automatica, vol. 18, pp.513-528.

Johnson C.R. (1980), Input matching error augmentation, self-tuning and output error identification : Algorithmic similarities in discrete adaptive model-following. IEEE Transactions on Automatic Control, vol. 25, pp. 697-703.

Johnson C.R. (1984), Adaptive IIR Filtering : Current results and open issues. IEEE Transactions on Information Theory, vol.IT-30, pp.237-250.

Kalman R.E. (1958), Design of a self-optimizing control systems. Transaction of the ASME, vol.80, pp. 468-478.

Kel'mans G.K., Poznyak A.S.,Chernitser A.V. (1981), Adaptive locally optimal control, International Journal of System Sciences, vol.12, pp.235-254.

De Keyser R.M., Van Canwenberghe G.R. (1983), Self-tuning prediction and Control. International Journal of Systems Sciences, vol.16, pp.351-366.

Kumar R. (1984), Simultaneous adaptive control and identification via weighted least-square algorithm. IEEE Transactions on Automatic Control, vol. AC-29, pp.259-263.

Kumar R., Moore J.B. (1982), Convergence of adaptive minimum variance controllers via weighting coefficient selection. IEEE Transactions on Automatic Control, vol.AC-27, pp.146-153.

Kushner H.J., Kumar R. (1982), Convergence and rate of convergence of a recursive identification and adaptive control method which uses truncated estimators . IEEE Transaction on Automatic Control, vol.27, pp. 775-782.

Landau I.D. (1974), A survey of model reference adaptive techniques - theory and applications. Automatica, vol. 10, pp.353-379.

Landau I.D. (1976), Unbiesed recursive identification using model reference techniques. IEEE Transactions on Automatic Control, vol.AC-21, pp.199-202.

Landau I.D. (1979), Adaptive Control. The Model Reference Approach , Marcel Dekker, Inc., New York.

Landau I.D. (1981a), Model reference adaptive controllers and stochastic self-tuning regulators - a unified approach. Journal of Dynamic Systems , Measurement and Control, vol. 103, pp.404-416.

Landau I.D. (1981b), Combining model reference adaptive controllers and stochastic self-tuning regulators. Automatica, vol.18, pp.77-84.

Landau I.D. (1982a), Martingale convergence analysis of adaptive schemes - a feedback approach. IEEE Transactions on Automatic Control, vol. AC.-27, pp.716-719.

Landau I.D. (1982b), Near supermartingales for convergence analysis of recursive identifications and adaptive control schemes. International Journal of Control, vol.35, pp.197-226.

Landau I.D. (1984), A feedback system approach to adaptive filtering. IEEE Transactions on Information Theory, vol. IT-30, pp.251-262.

Landau I.D., Lozano R. (1981), Redesign of explicit and implicit discrete time model reference adaptive control schemes. International Journal of Control, vol.33, pp.247-268.

Ljung L. (1977a), On positive real transfer functions and the convergence of some recursions. IEEE Transactions on Automatic Control, vol. 22, pp.539-551.

Ljung L. (1977b), Analysis of recursive stochastic algorithms. IEEE Transactions on Automatic Control, vol.AC-22, pp. 551-575.

Ljung L., Landau I.D. (1980), Model reference adaptive systems and self-tuning regulators - some connections. Proceedings 7th IFAC Congress, vol. 3, pp.1973-1980, Helsinki.

Ljung L., Soderström T. (1983), Theory and practice of recursive identification. The MIT Press Cambridge , Massachusets

Ljung L., Trulsson E. (1981), Adaptive control based on explicit criterion minimization. IFAC 8th Triennial World Congress, vol.VII, Kyoto, pp.1-6.

Monopoli R.V. (1974), Model reference adaptive control with an angmented error signal. IEEE Transactions on Automatic Control, vol. AC-19, pp.424-484.

Mudrov V.I., Kushko V.L. (1976), Measurement processing methods. Sovetskoye Radio, Moscow (in Russian).

Narendra K.S.(1980), Recent developments in adaptive control. In Unbehauen H. (Editor) (1980), Methods and Applications in Adaptive Control, Springer Verlag, Berlin, pp.84-101.

Narendra K.S., Lin Y.H. (1980), Stable discrete adaptive control. IEEE Transactions on Automatic Control, vol.25, pp.456-461.

Narendra K.S., Monopoli R.V. (Editors) (1980), Applications of adaptive control. Academic Press, New York.

Narendra K.S., Valavani L.S. (1978), Stable adaptive controller design - direct control. IEEE Transactions on Automatic Control, vol.AC-23, pp.570-583.

Narendra K.S., Valavani L.S. (1979), Direct and indirect adaptive control. Automatica, vol.15, pp.653-664.

Panuska V.(1968), A stochastic approximation method for identification of linear system using adaptive filtering. Proceedings Joint Automatic Conference, vol.2, pp.451-460.

Parks P.S. (1966), Liapunov redesign of model reference adaptive control systems. IEEE Transactions on Automatic Control, vol. AC-11, pp.362-367.

Peterka V.(1972), On steady state minimum variance control strategy. Kybernetica, v.8, pp.219-232.

Petrov B.N., Rutkovskiy V.Yu.,Zemlyakov S.D. (1980), Adaptive coordinate parametric control. Nauka, Moscow (in Russian).

Proceedings of the IEEE, Special Issue on Adaptive Systems,1976, vol. 64, No.8.

Rao C.R. (1965), Linear statistical inference and applications. Johns Wiley & Sons, Inc., New York.

Saridis G.H.(1974), Stochastic approximation methods for identification and control - a survey . IEEE Transactions on Automatic Control, vol. 19, pp.798-809.

Saridis G.N. (1977), Self-organizing control of stochastic systems. Marcel Dekker, New York.

Sin K.S., Goodwin G.S. (1982), Stochastic adaptive control using a modified least squares algorithm. Automatica, vol.18, pp.315-321.

Solo V.(1979), The convergence of AML. IEEE Transactions on Automatic Control, vol.24, pp.958-963.

Sragovich V.G. (1976), Theory of adaptive systems. Nauka, Moscow, (in Russian).

Sragovich V.G. (1981), Adaptive control. Nauka, Moscow (in Russian).

Stormer R.R. (1959), Adaptive and self-optimizing control systems. IRE Transactions on Automatic Control, vol.3, pp.65-68.

Trulsson E. (1983), Adaptive control based on explicit criterion minimization. Dissertation No. 106, Linköping Studies in Science and Technology.

Tsypkin Ya.Z. (1971), Adaptation and learning in automatic systems.Academic Press, New York.

Tsypkin Ya.Z. (1983), Optimality in identification of linear plants, Int. J. Systems Sci., 1983, vol. 14, No.1, pp. 59-74.

Tsypkin Ya.Z. (1984), Fundamentals of information theoretical identification. Nauka, Moscow (in Russian).

Unbehauen H. (Editor) (1980), Methods and applications in adaptive control. Proceedings of an International Symposium, Bochum, 1980. Springer Verlag, Berlin-Heidelberg-New York.

Unbehauen H., Schmidt Ch.R. (1975), Status and industrial application of adaptive control systems. Automatic Control Theory and Application, vol.3, pp.1-12.

Weber W. (1971), Adaptive regelungs systeme I,II,. R.Oldenbourg Verlag, München - Wien.

Wellstead P.E., Edmunds J.M., Prager D.L., Zanker P. (1980), Classical and optimal self-tuning control. International Journal of Control, vol.31, pp.1003-1005.

Wellstead P.E., Sanoff S.P. (1981), Extended self-tuning algorithm. International Journal of Control, v.34, pp.433-455.

Wieslander J., Wittenmark B. (1971), An approach to adaptive control using real time identification. Automatica, vol.7, pp.211-217.

Wittenmark B. (1975), Stochastic adaptive control methods (survey). International Journal of Control, vol. 21, pp.705-730.

Young P.C. (1966), Process parameter estimation and self-adaptive control. In Hammond P.H. (Editor) (1966), Theory of self-adapting control system, Plenum Press, New York.

Young P.C., Hastings-James R. (1981), Identification and control of discrete linear systems. Subject to disturbances with rational spectral density. International Journal of Control, vol. 34, pp.98-121.

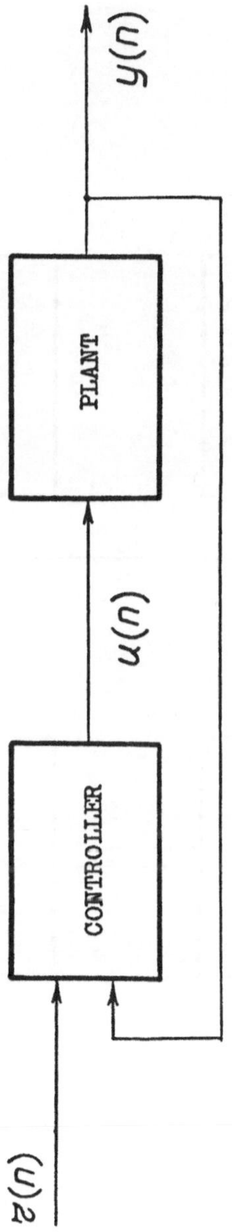

Fig. 1

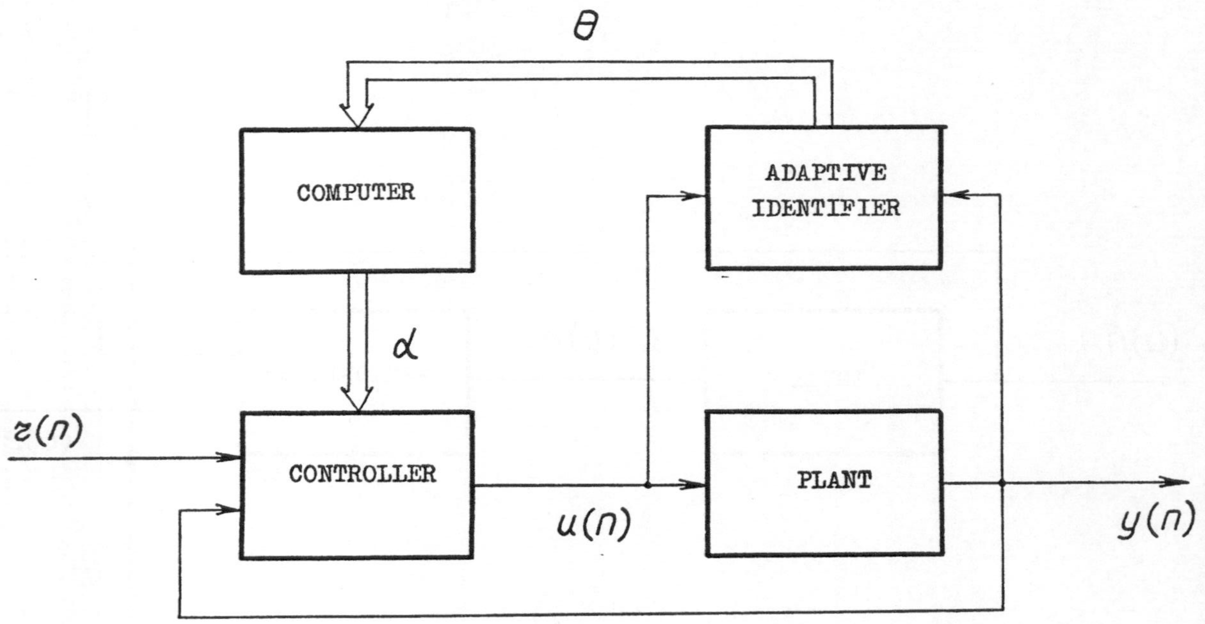

Fig. 2

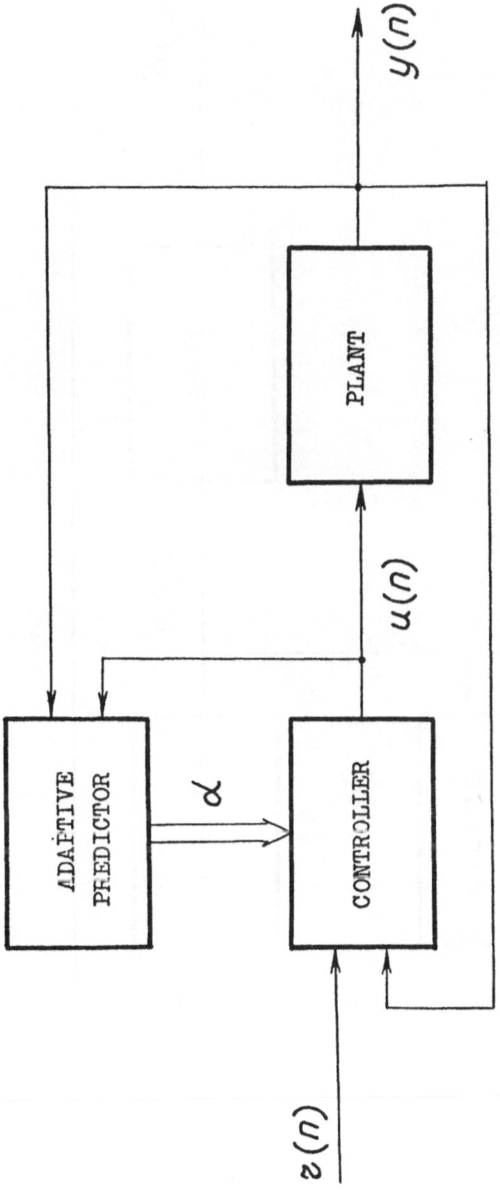

Fig. 3

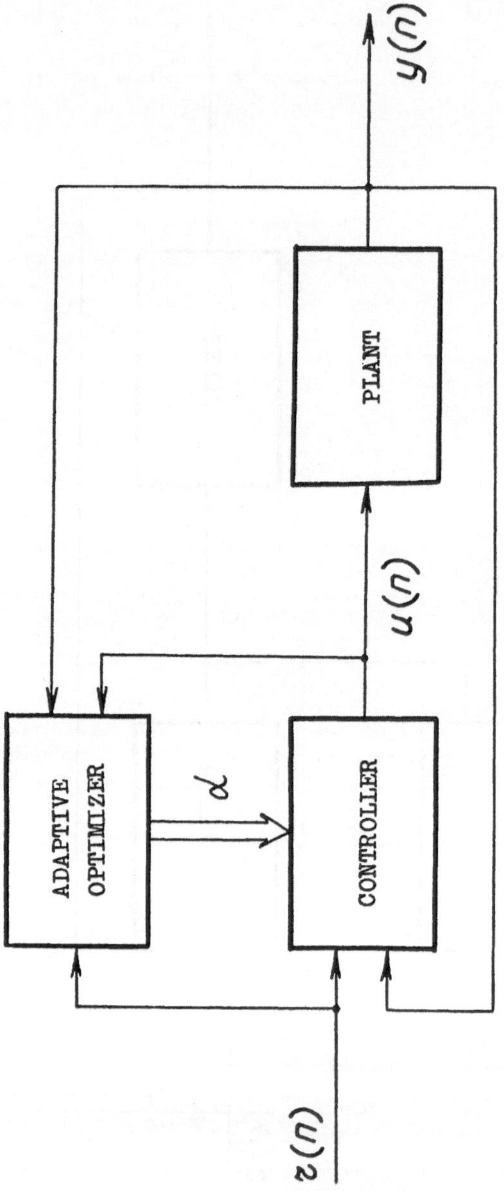

Fig. 4

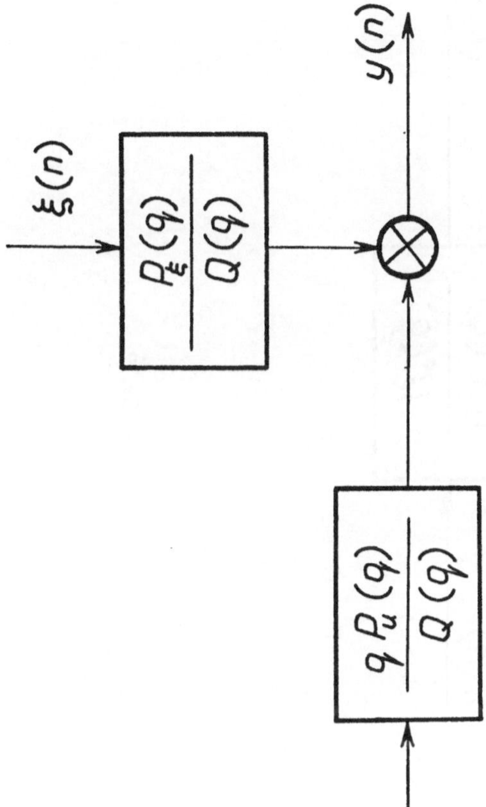

Fig. 5

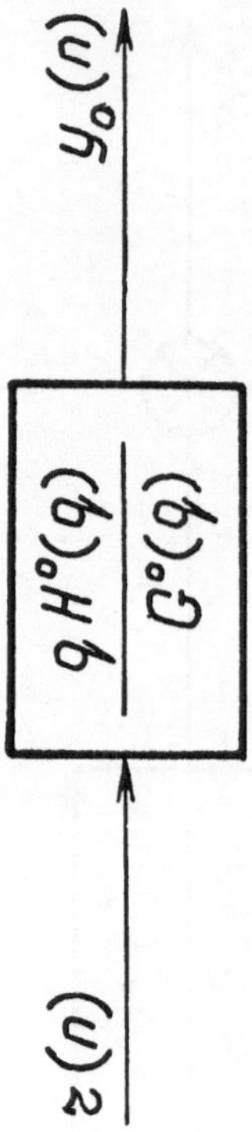

Fig. 6

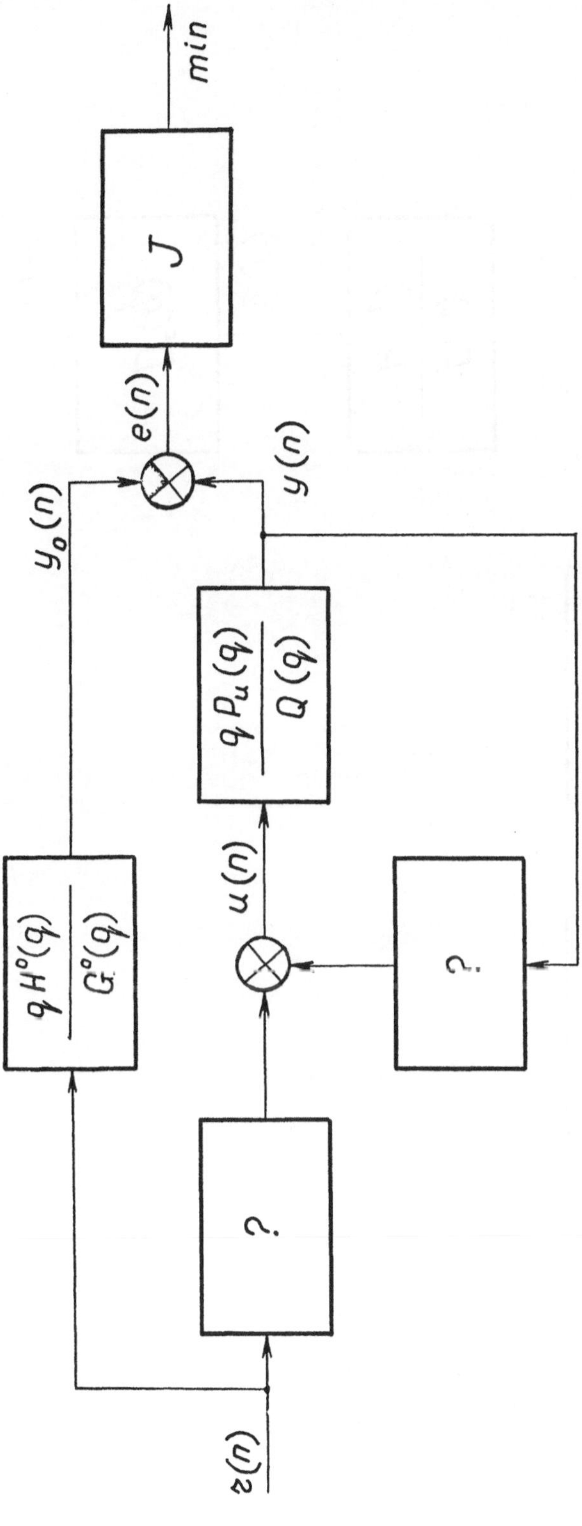

Fig. 7

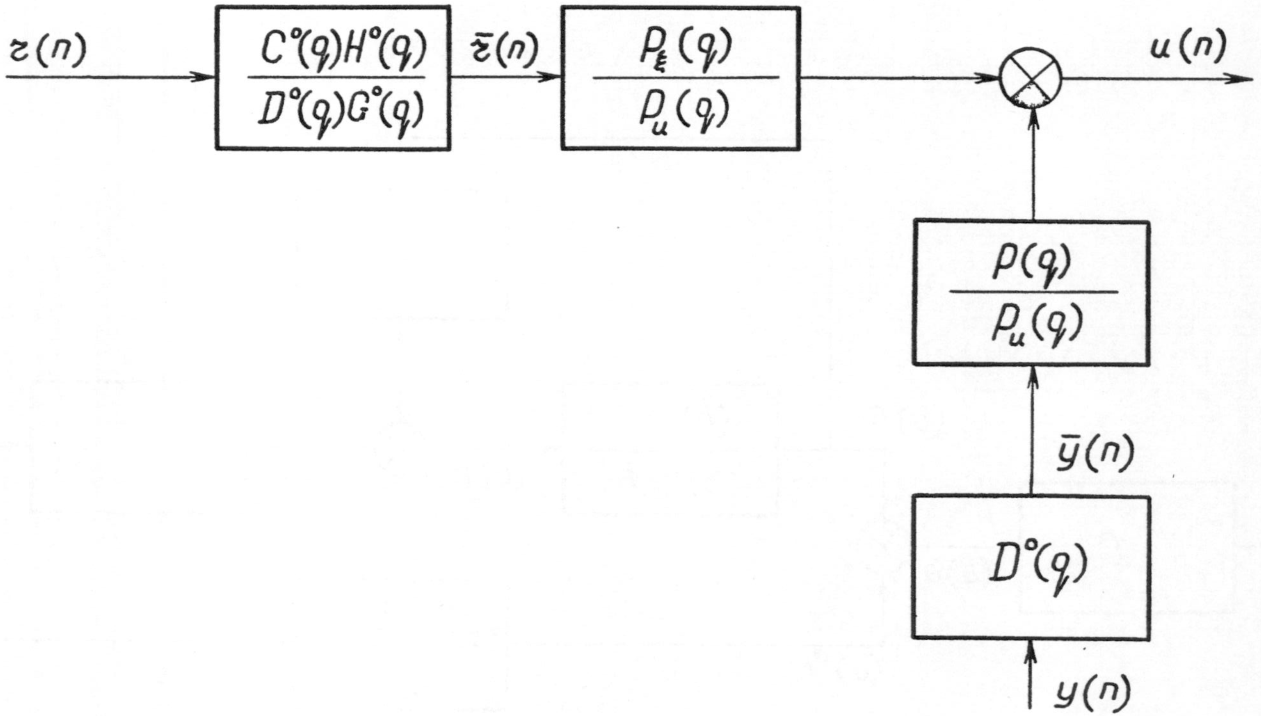

Fig. 8

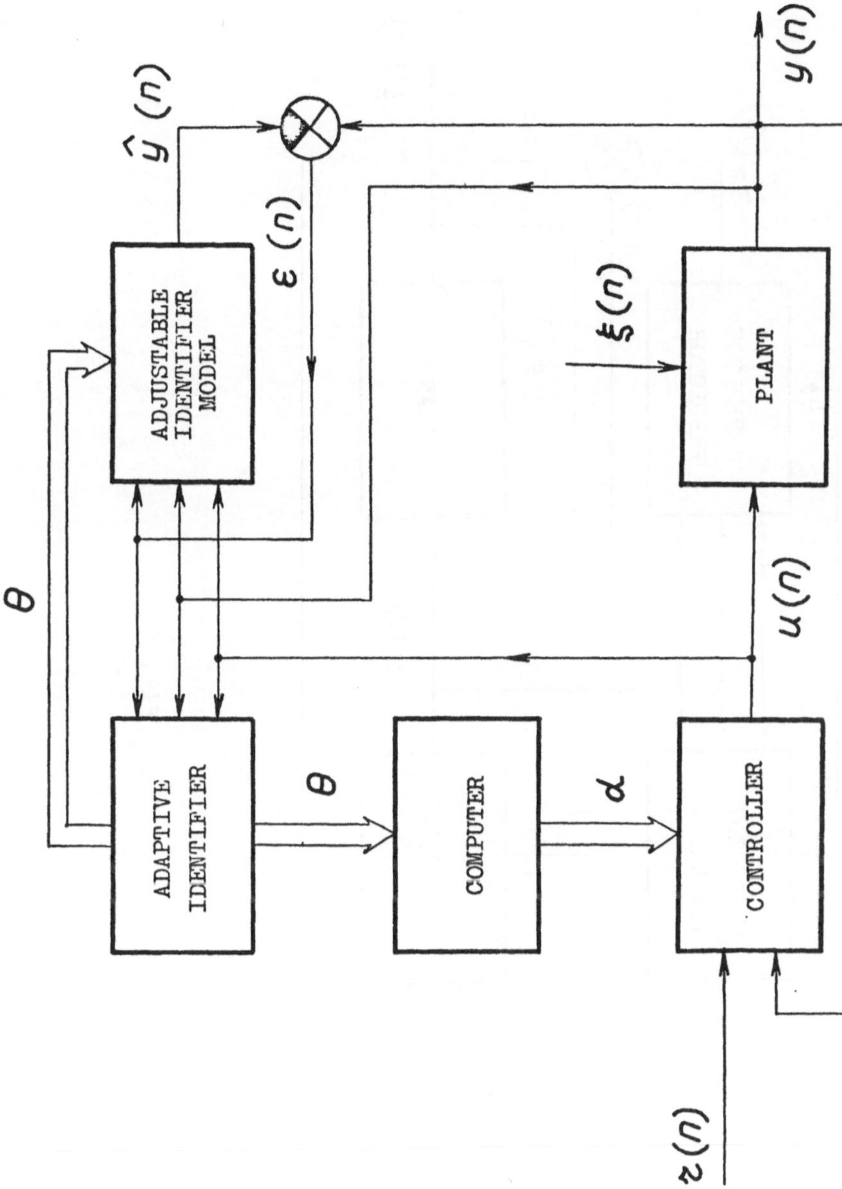
ADJUSTABLE IDENTIFIER MODEL
ADAPTIVE IDENTIFIER
COMPUTER
CONTROLLER
PLANT
$\hat{y}(n)$
$\varepsilon(n)$
$\xi(n)$
$y(n)$
$u(n)$
$r(n)$
θ
θ
α

Fig. 9

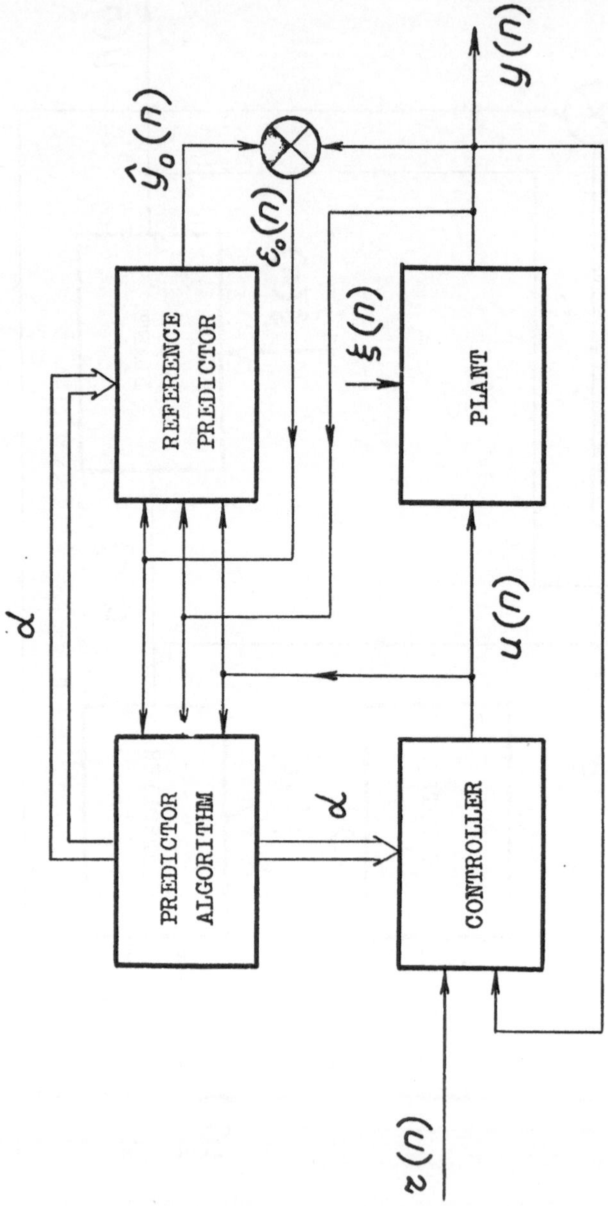
REFERENCE PREDICTOR
PLANT
PREDICTOR ALGORITHM
CONTROLLER
$\hat{y}_o(n)$
$y(n)$
$\varepsilon_o(n)$
$\xi(n)$
$u(n)$
$z(n)$
α
α

Fig. 10

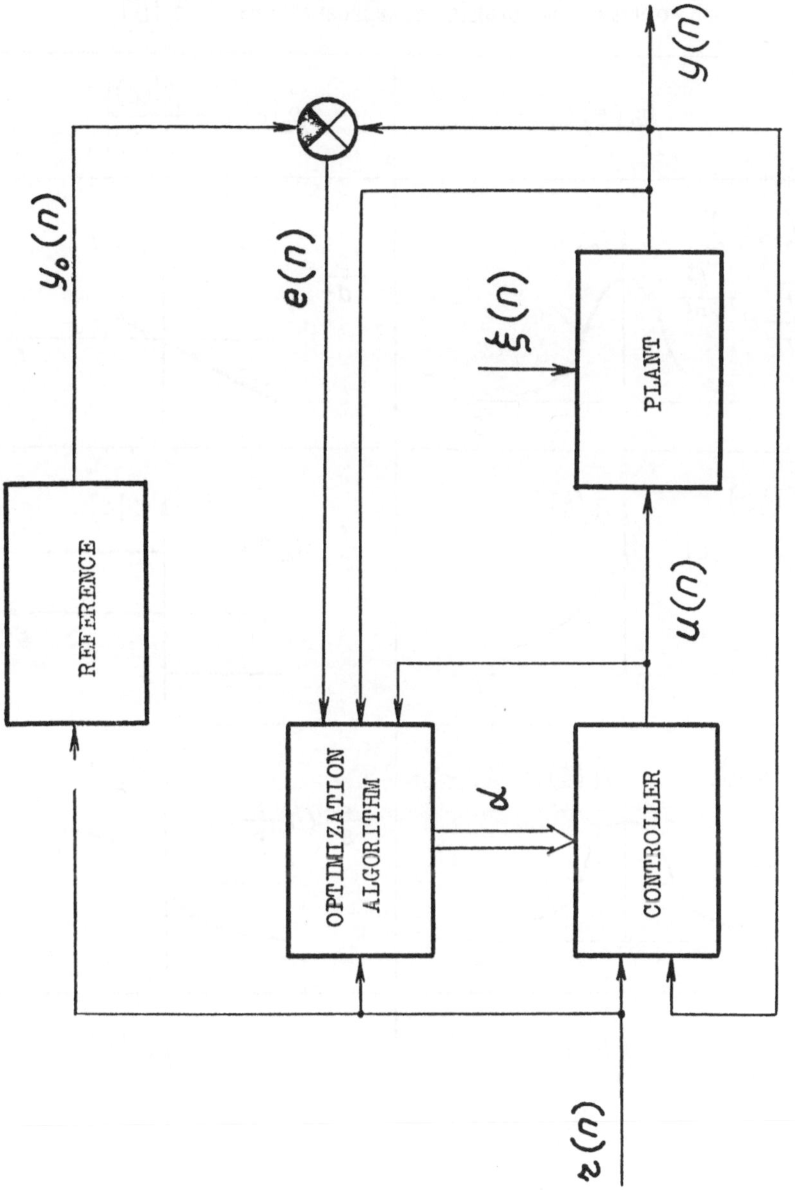
REFERENCE
OPTIMIZATION ALGORITHM
CONTROLLER
PLANT
$y_o(n)$
$e(n)$
$\xi(n)$
$u(n)$
$y(n)$
$r(n)$
α

Fig. 11

TABLE 1

OPTIMAL NONLINEAR TRANSFORMATIONS $\varphi_o[\varepsilon]$

№	$p_o(\xi)$	$\varphi_o[\varepsilon] = -\left.\frac{p_o'(\xi)}{p_o(\xi)}\right\vert_{\xi=\varepsilon}$
1	$N(0,\sigma^2)$ $\frac{1}{\sqrt{2\pi}\sigma} e^{-\frac{\xi^2}{2\sigma^2}}$ 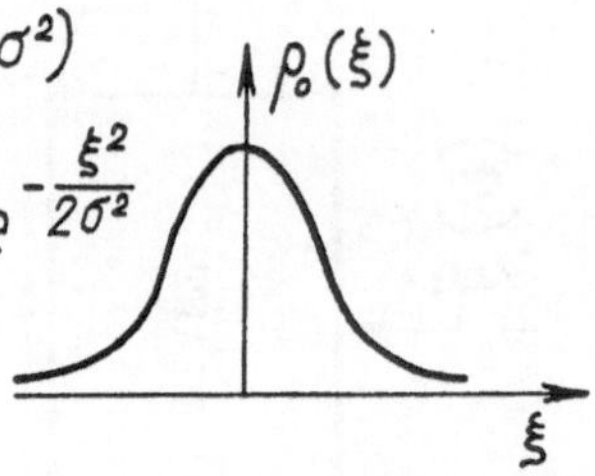	$\frac{\varepsilon}{\sigma^2}$ 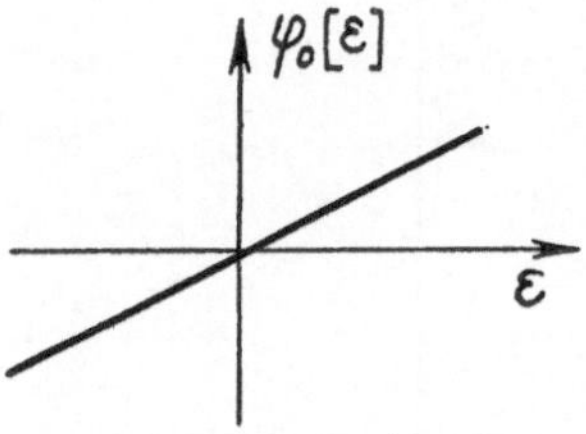
2	$L(0,s)$ $\frac{1}{2s} e^{-\frac{\lvert\xi\rvert}{s}}$ 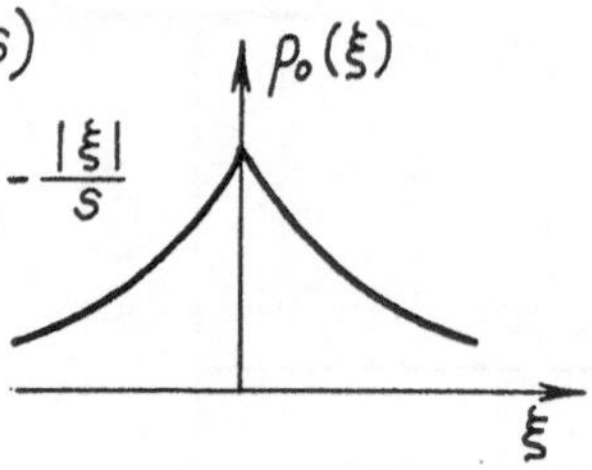	$\frac{1}{s}\,\mathrm{sign}\,\varepsilon$ 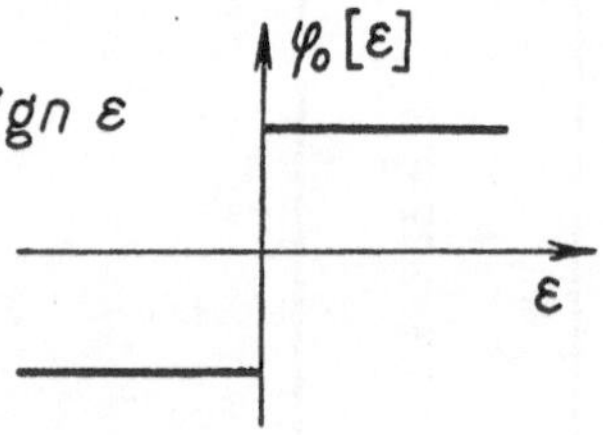
3	$Se(0,s)$ $\frac{1}{2s}\,\mathrm{sech}^2\frac{\xi}{s}$ 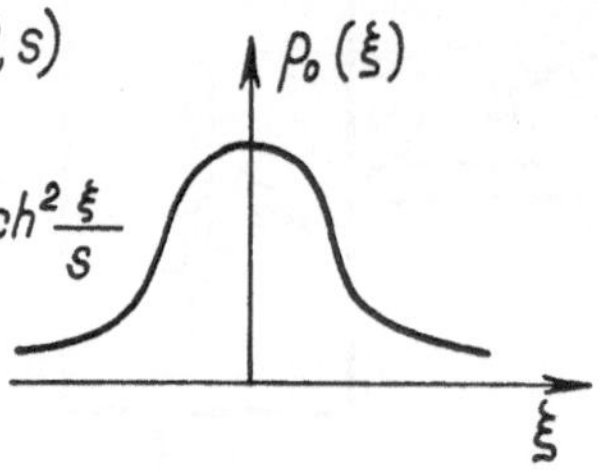	$\frac{2}{s}\,\mathrm{th}\,\frac{\varepsilon}{s}$ 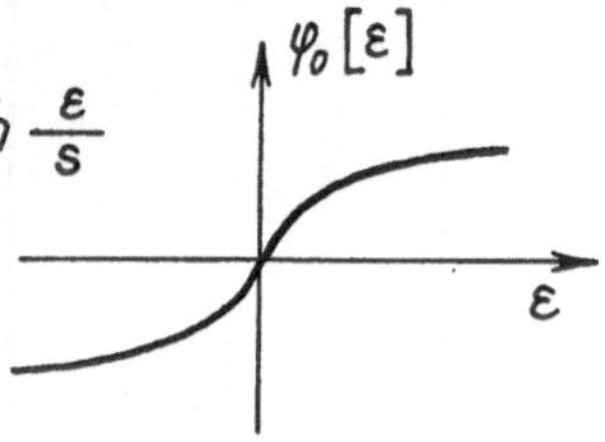
4	$C(0,s)$ $\frac{s}{\pi}\,\frac{1}{s^2+\xi^2}$ 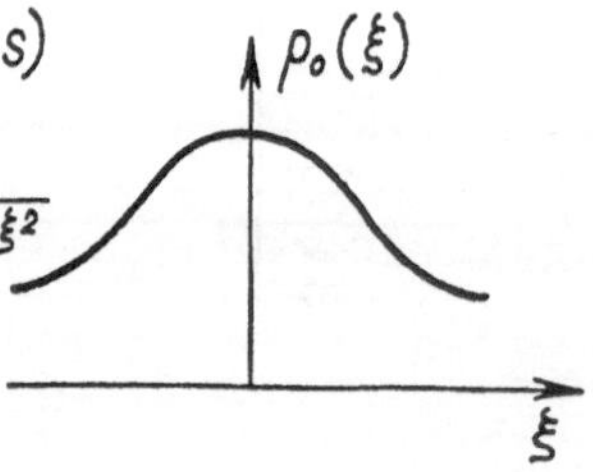	$\frac{2\varepsilon}{s^2+\varepsilon^2}$ 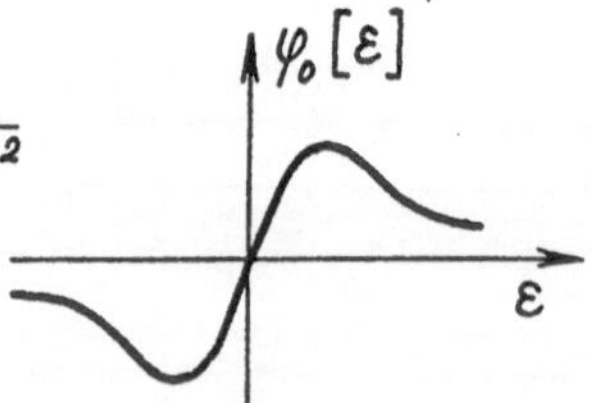

TABLE 2

ARGUMENTALLY OPTIMAL ON A CLASS NONLINEAR TRANSFORMATIONS $\varphi_*[\varepsilon]$

CLASS $\mathcal{P}$	$p_*(\xi)$	$\varphi_*[\varepsilon] = -\left.\frac{p_*'(\xi)}{p_*(\xi)}\right\vert_{\xi=\varepsilon}$
$\mathcal{P}_1 : p_0(0) \geqslant \frac{1}{2s_1}$		
s_1 IS SMALL	$\frac{1}{2s_1} e^{-\frac{\lvert\xi\rvert}{s_1}}$	$\frac{1}{s_1}\, \mathrm{sign}\, \varepsilon$
	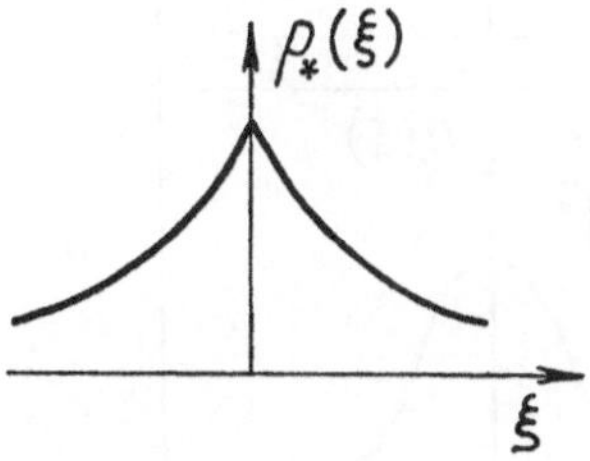	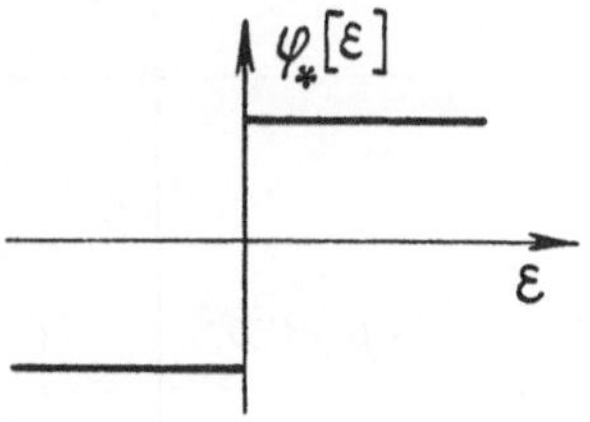
s_1 IS LARGE	$C e^{-\lambda[\lvert\xi\rvert\sqrt{\alpha\xi^2+\beta} + \frac{\beta}{\sqrt{\alpha}} \mathrm{Arch}\sqrt{\frac{\alpha}{\beta}}\xi]}$	$\lambda[\frac{2\alpha\varepsilon^2+\beta}{\sqrt{\alpha\varepsilon^2+\beta}}\, \mathrm{sign}\, \varepsilon + \sqrt{\beta}\, \mathrm{Arch}\sqrt{\frac{\alpha}{\beta}}\varepsilon]$
	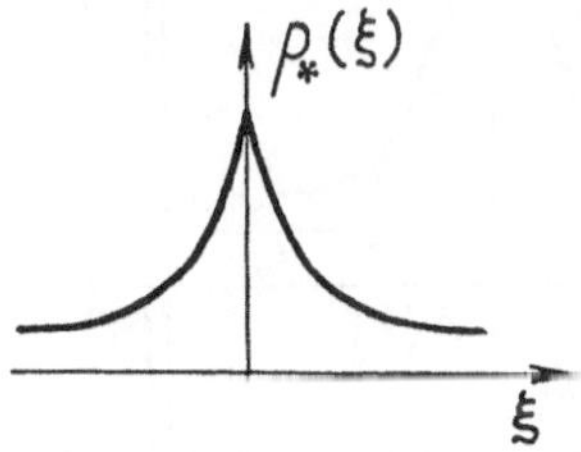	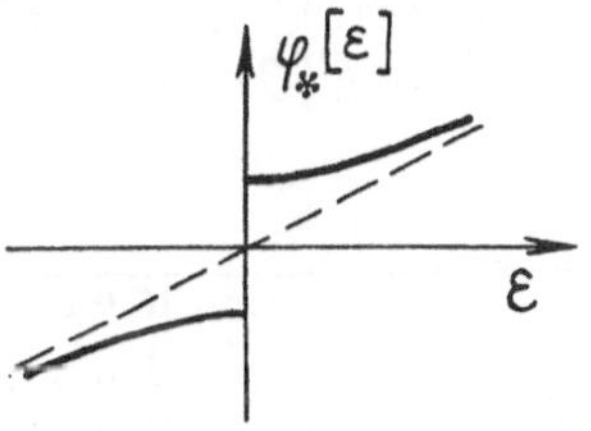
$\mathcal{P}_2 : \sigma^2(p_0) \leqslant \sigma_1^2$	$\frac{1}{\sqrt{2\pi}\,\sigma_1} e^{-\frac{\xi^2}{2\sigma_1^2}}$	$\frac{\varepsilon}{\sigma_1^2}$
	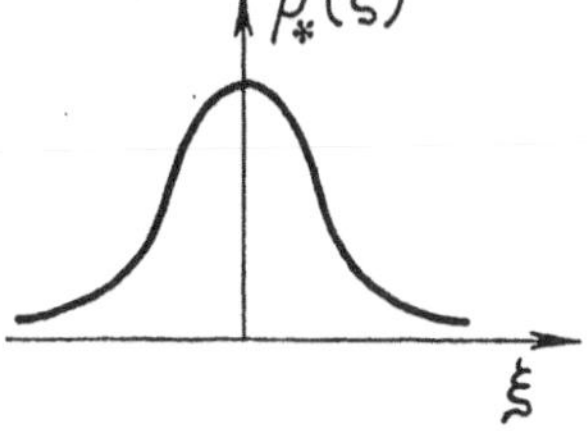	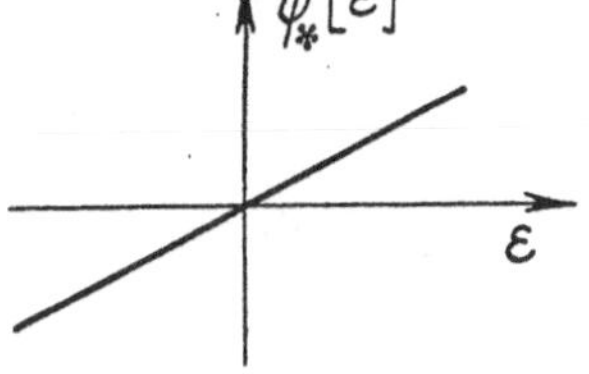

TABLE 3

CRITERIALLY OPTIMAL ON A CLASS NONLINEAR TRANSFORMATIONS $\varphi_*[\varepsilon]$

CLASS $\mathcal{P}$	$p_*(\xi)$	$\varphi_*[\varepsilon] = -\left.\frac{p_*'(\xi)}{p_*(\xi)}\right\vert_{\xi=\varepsilon}$
$\mathcal{P}_1: p_0(0) \geq \frac{1}{2s_1} > 0$	$\frac{1}{2s_1} e^{-\frac{\vert\xi\vert}{s_1}}$ 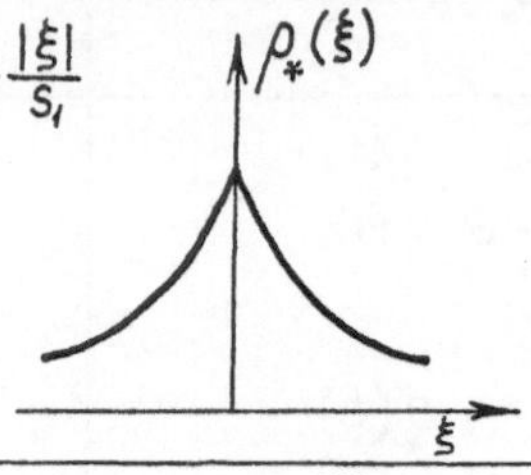	$\frac{1}{s_1}\,\mathrm{sign}\,\varepsilon$ 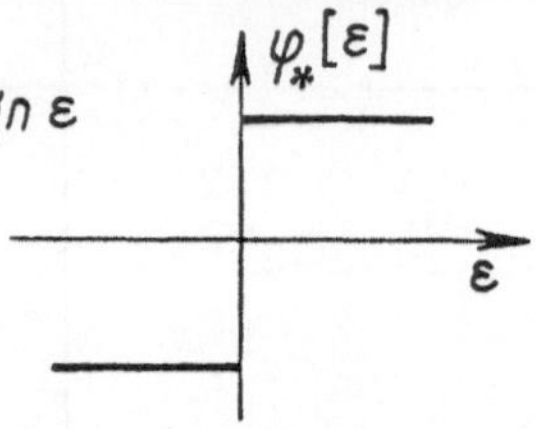
$\mathcal{P}_2: \sigma^2(p_0) \leq \sigma_1^2$	$\frac{1}{\sqrt{2\pi}\sigma_1} e^{-\frac{\xi^2}{2\sigma_1^2}}$ 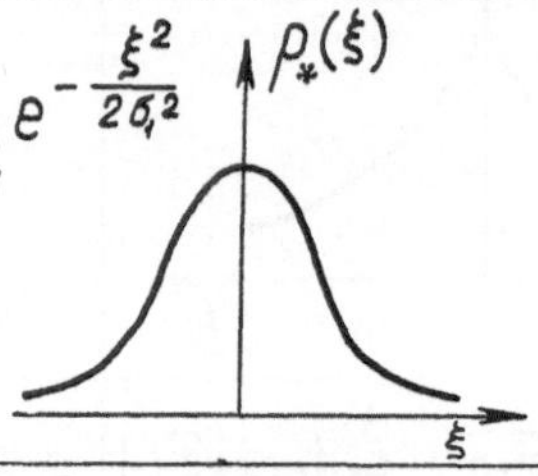	$\frac{\varepsilon}{\sigma_1^2}$ 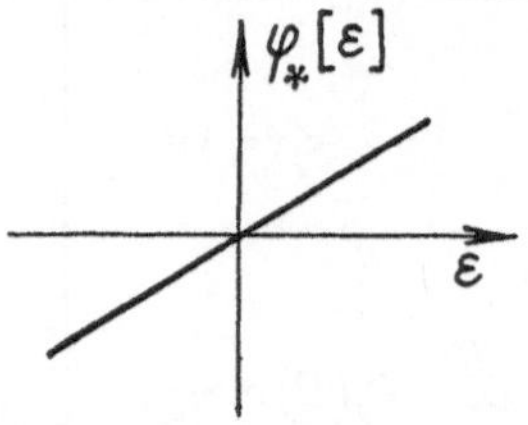
$\mathcal{P}_3: p_0(\xi) \geq (1-\alpha) \times N(0,\sigma_N^2)$	$\begin{cases} \frac{1-\alpha}{\sqrt{2\pi}\sigma_N} e^{-\frac{\xi^2}{2\sigma_N^2}}, & \vert\xi\vert \leq \Delta \\ \frac{1-\alpha}{\sqrt{2\pi}\sigma_N} e^{-\frac{\Delta}{\sigma_N^2}(\vert\xi\vert - \frac{\Delta}{2})}, & \vert\xi\vert > \Delta \end{cases}$ 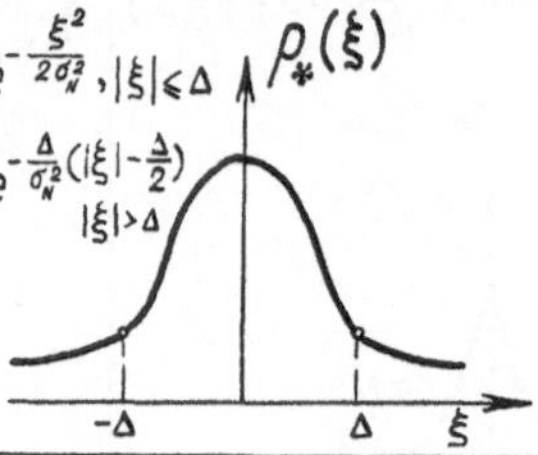	$\begin{cases} \frac{\varepsilon}{\sigma_N^2}, & \vert\varepsilon\vert \leq \Delta \\ \frac{\Delta}{\sigma_N^2}\,\mathrm{sign}\,\varepsilon, & \vert\varepsilon\vert > \Delta \end{cases}$ 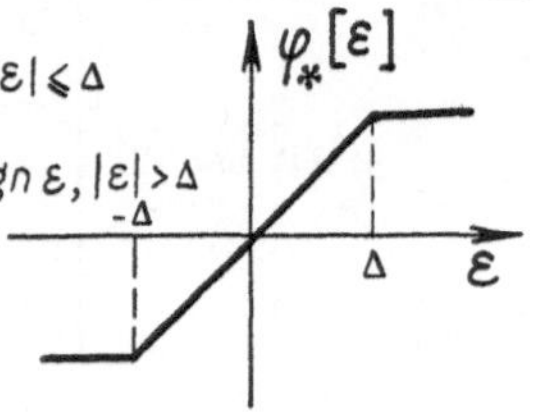
$\mathcal{P}_4: p_0(\xi) \geq (1-\alpha) \times R(0,l_R)$	$\begin{cases} \frac{1-\alpha}{2\Delta}, & \vert\xi\vert \leq \Delta \\ \frac{1-\alpha}{2\Delta} e^{-\frac{1-\alpha}{\alpha\Delta}(\vert\xi\vert - \Delta)}, & \vert\xi\vert > \Delta \end{cases}$ 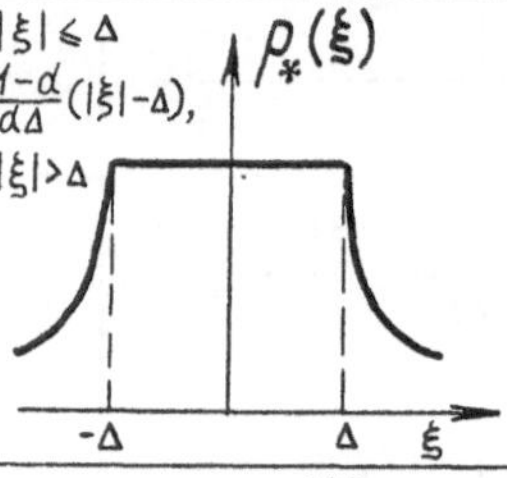	$\begin{cases} 0, & \vert\varepsilon\vert \leq \Delta \\ \frac{1-\alpha}{\alpha\Delta}\,\mathrm{sign}\,\varepsilon, & \vert\varepsilon\vert > \Delta \end{cases}$ 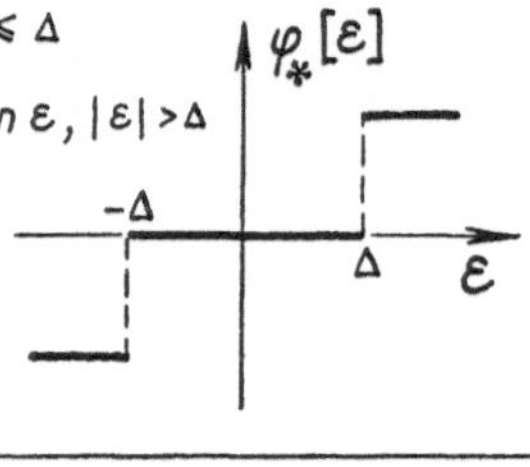
$\mathcal{P}_5: \int_{-l}^{l} p_0(\xi)\,d\xi = 1$	$\frac{1}{l}\cos^2\frac{\pi\xi}{2l}$ 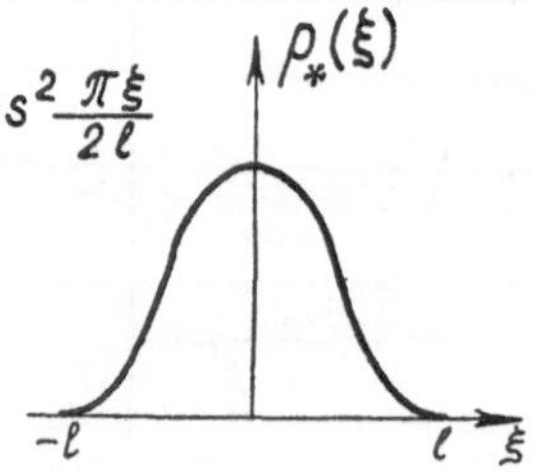	$\frac{\pi}{l}\,\mathrm{tg}\,\frac{\pi\varepsilon}{2l}$ 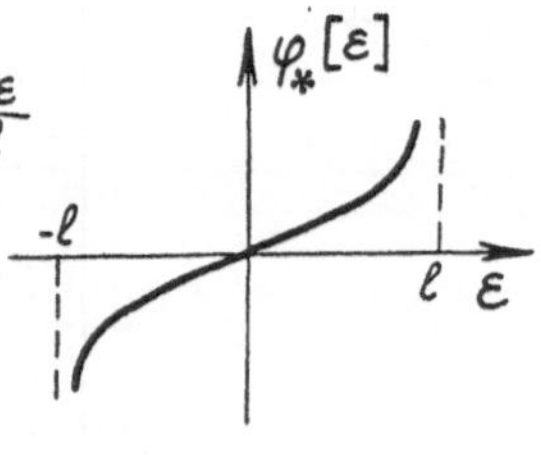

INTELLIGENT CONTROL-OPERATING SYSTEMS IN UNCERTAIN ENVIRONMENTS

by

George N. Saridis

Electrical, Computer, and Systems Engineering Department
Rensselaer Polytechnic Institute
Troy, New York 12180-3590

ABSTRACT

Systems operating in uncertain environments with minimum interaction with a human operator, may be managed by controls with special considerations. They result in a Hierarchically Intelligent Control System, the higher levels of which may be modeled as knowledge based system processing various types of information with Entropy as an analytic measure. The concept of Entropy of statistical Thermodynamics, is used to express the average value of the performance criterion of a feedback control, encountered at the lower levels of the system. Thus, the resulting optimal control problem may be recast as an information theoretic one, which minimizes the entropy of selecting the feedback controls. This unifies the treatment of all the levels of a Hierarchically Intelligent Control System by a mathematical programming algorithm which minimizes the sum of their entropies. The resulting "Intelligent Machine" is composed of three levels hierarchically ordered in decreasing intelligence with increasing precision: the organization level, performing information processing tasks like planning, decision making, learning and storage and retrieval of information from a long-term memory; the coordination level, dealing again with information processing tasks like learning, lower level decision making and dealing with short-term memory only and the control level, which performs the execution of various tasks through hardware using feedback control methods.

1. INTRODUCTION

Modern technology has opened new horizons of exploration, hardly thought before as possible by a human mind. It may find applications from the unstructured environments of deep space all the way to the structured assembly lines of the "factories of the future". All these diverse environments have in common the predominance of uncertainties about their configuration, mode of operation, hazards as well as appropriate strategies for their exploration. In view of the limitations of the human body and mind to guide and control activities in such uncertain environments it is highly desirable to devise machines that could perform the same tasks with minimum interaction with the human operator. These machines are called "Intelligent Machines" and should be equipped with intelligence and dexterity resembling those of the human being.

Most attempts made to create such intelligent machines are based on digital computers which utilize human logic and heuristics in an effort to imitate the function and the activities of the human brain, Albus (1975), Ashby (1965). In spite of the tremendous capacity of the modern computers to heuristically organize and execute complex tasks, it is highly desirable to generate an analytic approach to provide a means to optimize the operation and thus produce a much more efficient machine.

In an attempt to formulate an analytic approach to design an intelligent machine, one may first consider the diversity of functions involved in such an operation, from the computer organization and coordination of tasks to their hardware execution. Such functions have been submitted to analytical treatment within the realm of their own individual disciplines with very little consideration for cross disciplinary integration, Hayes-Roth, et. al. (1983). They should involve instead interaction of several disciplines like Control, Operations Research, and Artificial Intelligence, (see Figure 1), as suggested by Saridis (1983). The main obstacle for such a development is the missing of an analytic measure common to all those functions that its aggregated value may represent the "performance cost" of the whole process.

The generic feature for the analytic design of intelligent machines is the uncertainties encountered in their environment of operation as well as the decision making of the machine itself. The problem of uncertainty has been dealt with before in scientific literature and has been treated probabilistically. Such a point of view has been taken for the design of intelligent machines and is discussed in the next sections, Saridis (1984).

2. UNCERTAINTY AND ITS MEASURE

Uncertainty is an old commodity in science and mathematics. It has been treated most of the time probabilistically by assigning a distribution function to describe how uncertain one is about the outcome of a particular event selected from a class of many possible events, and is discussed in Prigogine (1980). Events investigated have been selected from games of chance, T. Bayes (1763), Statistical Thermodynamics, Boltzmann (1905), quantum mechanics, Heisenberg (1975), biological systems, Prigogine (1980), information theory, Shannon (1963), etc. They all assumed that the uncertainty was caused by the poor or inefficient modeling of the environment and a prediction of the outcome could be best modeled in a space of randomized events. In all these cases Entropy was used as a measure of uncertainty.

Decision making has introduced a different type of probabilities which deals with the uncertainties of the selection of a certain event based on the knowledge of the system. Such probabilities can be modified with the experience of the system, and are called subjective probabilities. Saridis (1977) coı ·dered such probabilities in conjunction with reducible uncertainties in a Self-Organizing Control System. The other kind of probabilities deal only with irreducible uncertainties of complex environmental models and are called objective probabilites. Both probabilities can be treated in the classical manner and utilize Entropy as a measure of uncertainty in a way similar to the one that Boltzmann suggested for statistical thermodynamics.

There is, though, a difference in the behavior of the entropies of subjective and objective probabilities. It can be easily shown that the entropy of subjective probabilities is reduced with learning; e.g., when with acquisition of new information about the event the probability density becomes less flat. The objective probabilities do not experience learning and their entrc y remains unchanged. The theory of intelligent machines deals mainly with subjective probabilities that require decision making at various levels and they are subject to sequential improvement which is defined as learning.

It has been argued by many, that subjective probabilities when dealing with human judgement and decision making may become, under relaxed conditions, measures of belief of the outcome of uncertain events and therefore apply to intelligent beings. This argument led to a more general Mathematical Theory of Evidence, Shafer (1976) of which Probability Theory is a special case. Since the purpose of this work is to develop a theory of intelligent machines not necessarily possessing any human judgement, the probabilistic model is sufficient for the purpose. However, generalizations are possible, see Stephanou and Lu (1983).

Boltzmann's theory states that the entropy of a gas changing states isothermally at temperature T is given by:

$$S = -k \int_{\Omega_x} \frac{\psi - H}{kT} e^{\frac{\psi - H}{kT}} dx \quad (1)$$

where ψ is Gibbs energy, $\psi = - kT \log \int e^{\frac{-H}{kT}} dx$, H is the total energy of the system and k is Boltzmann's constant. Due to the size of the problem and the uncertainties involved in describing its dynamic behavior a probabilistic model was assumed where the entropy is a measure of the uncertainty of molecular distribution. If p(x) is defined as the probability of a molecule being in state x, thus assuming

$$p(x) = \exp \{ \frac{\psi - H}{kT} \} \quad (2)$$

where p(x) must satisfy the "incompressibility" property in the state space, Ω_x,

$$\frac{dp}{dt} = 0 \quad (3)$$

The incompressibility property of the probability density function is a differential constraint, which in the case of perfect gases yields Liouville's equation. Substituting (2) into (1), the entropy of the system becomes

$$S = -k \int_{\Omega_x} p(x) \ln p(x) \, dx \quad (4)$$

In the 1940's, Shannon (1963) resolved the problem of measuring information by defining (negative) entropy as a measure of the uncertainty of transmission of information:

$$H = - \int_{\Omega_s} p(s) \ln p(s) \, ds \quad (5)$$

where Ω_s is the space of information signals transmitted.

Equation (4) is of the same form as eq. (5) used to describe entropy of an information source when the state space of the molecules of a perfect gas Ω_x is replaced by the space of information signals Ω_s. In both cases the entropy thus defined is a measure of uncertainty about the state of the system, obtained as a function of a probability density exponential function of the involved system energy.

The similarities of the two formulations are obvious and they can be extended to cover other types of systems where the states of the system are assumed to be exponentially distributed because of uncertainties in observations. This observation shall lead to the development of intelligent control systems where high level decision can be modeled by an information theoretic system described by distributions of subjective probabilities with their appropriate entropies. The control problem at the hardware execution level can be also solved probabilistically using a model derived from Boltzmann's statistical thermodynamics with its entropy as the cost function. Thus entropy provides a common measure of all the levels of an intelligent machine.

3. HIERARCHICALLY INTELLIGENT CONTROL FOR INTELLIGENT MACHINES

Intelligent Machines are hierarchical structures in the order of intelligence and inverse order of precision. They differ from other hierarchical structures described in Findeisen, Baily, et. al. (1980) in this particular type of ordering, putting emphasis on the intelligence of the machine. An intelligent machine, designed according to the principle of increasing precision and decreasing intelligence should contain three levels, defined by Saridis (1983). This principle should be interpreted as a constraint of the intelligent system while the terms intelligence and precision both related to the removal of uncertainty at different levels are clarified below:

- Intelligence is measured by the rate of knowledge flowing at the highest levels.
- Precision is the function of removing the uncertainty in the execution of the various tasks, at the lowest levels.

The three levels are:

i. The organization level, which should perform such operations as planning, high level decision making, learning and data storage and retrieval from long term memories. It may require high level information processing like the knowledge based systems encountered in Artificial Intelligence, Whinston (1977), will require large quantities of knowledge processing with little or no precision.

Its performance can be expressed in terms of entropies and entropy rates effectively.

ii. The coordination level, is an intermediate structure serving as an interface between the organization and execution level. It will perform operations as decision making and learning but with the use of a short term memory only. Such information processing can be accomplished by sequentially updating algorithms like decision schemata developed by Saridis and Graham (1984). Obviously, coordination involves less knowledge-processing than organization and the procedure is straight forward when one uses the subjective probabilities generated by the decision schemata, Saridis and Blumberg (1983), to evaluate the appropriate entropies. The precision levels are still minimal but they may be expressed in terms of the uncertainties of selecting the proper control.

iii. The hardware execution level, which is the lowest one in the hierarchy and is evaluated by the cost of executing the appropriate control functions. In Section 4 of this paper, this performance measure will be expressed as an entropy, thus completing the functions of an "intelligent machine" to be evaluated by entropies. Precision expressed in terms of certainty in execution comes nicely into the picture.

It will be shown that all the levels of a hierarchical intelligent control can be measured by entropies and their rates. Then the optimal operation of an "intelligent machine" can be obtained through the solution of the following mathematical programming problem.

The theory of intelligent machines may be postulated as the mathematical problem of finding the right sequence of decisions and controls for a system structured according to the principle of increasing precision with decreasing intelligence (constraint) such that it minimizes its total entropy.

The above analytic formulation of the "intelligent machine" problem as a hierarchically intelligent control problem was based on the use of entropy as a measure of performance at all the levels of the hierarchy. It has many advantages because of the tree-like structure of the decision making process, and brings together functions that belong to a variety of disciplines. Developing the entropy measures in the different parts of the intelligent machine is the subject of the next sections.

4. THE THEORY OF INTELLIGENT MACHINES-THE ORGANIZATION LEVEL OF CONTROL

It has been stipulated that learning, knowledge acquisition, planning, decision making and data storage and retrieval from long-term or short-term memories, are characteristic functions of the intelligent part of an "intelligent machine". Since all these functions are directly related to information processing, it is natural to assume that they contribute to the production of entropy in the information theoretic sense of Shannon (1963). Therefore, entropy may serve as a measure of the intelligent function system and its rate may describe the amount of knowledge processed in a way similar to the generalized law of information rates by Conant (1976).

The functions involved in the upper two levels of an intelligent machine are imitating functions of human behavior and may be treated as elements of knowledge-based systems, as in Hayes-Roth, et. al. (1983). Actually, the activities of planning, decision making, learning, data storage and retrieval, task coordination, etc. may be thought of as knowledge (information) handling and management. Therefore, the flow of knowledge is the main variable in the function of an intelligent machine.

- Knowledge is the function of removing the ignorance or uncertainty in the operation of an intelligent machine and may be measured by entropy, which is a measure of uncertainty.
- The Rate of Knowledge F is related to the flow of knowledge in the machine, has the ability to reduce uncertainty and it is a measure of its Intelligence.

It is measured in terms of entropy rates and must satisfy the generalized law of information rates,

$$F = F_T(x,y) + F_B(x,y) + F_C(x,y) + F_D(x,y) + F_N(x,y) \qquad (6)$$

where x are the inputs and y are the states of the machine and F symbolizes the total rate and

F_T = Throughput Rate

F_B = Blockage Rate

F_C = Coordination Rate

F_D = Internal Decision Rate

F_N = Noise Rate

Knowledge flow in a knowledge-based system is composed of

1. Knowledge Representation
2. Reasoning
3. Cognition
4. Languages

In an intelligent machine's organization level knowledge flow represents respectively,

1. Data Handling and Management performed through the computer memory
2. Planning and Decision performed by the CPU
3. Sensing and Data Acquisition obtained through I/O's
4. Formal Languages which define the software.

Subjective probabilistic models are assigned to the individual function and their entropies may be evaluated for every task that is executed, thus providing an analytic measure of the total activity.

5. THE THEORY OF INTELLIGENT MACHINES-THE COORDINATION LEVEL

The coordination level is involved with coordination, decision making and learning on a short term memory e.g., a buffer. It may utilize linguistic decision schemata with learning capabilities defined in Saridis and Graham (1982), and assign subjective probabilities for each action. The respective entropies may be obtained directly from these subjective probabilities.

A decision schema is a software device that maps a string from an input language $L(G_i)$ to each possible string belonging to one or more output languages $L(G_{oj})$

$$L(G_i) = \left\{\begin{matrix} x_1 \\ . \\ . \\ . \\ x_n \end{matrix}\right\} \xrightarrow{P} \bigcup_{j=1}^{\ell} L(G_{oj}) = \left\{\begin{matrix} y_{11} \\ . \\ . \\ . \\ y_{1m_j} \end{matrix}\right\} \cdots \left\{\begin{matrix} y_{\ell 1} \\ . \\ . \\ . \\ y_{\ell m_\ell} \end{matrix}\right\} \tag{7}$$

To each mapping there is associated a subjective probability $P \varepsilon \{p_{ijk}$, $i = 1,\dots n$, $j = 1,\dots l$, $k = 1,\dots m\}$ which may be used to select the proper output string related to the particular input string. A rule for a unique association is obtained for a given input by defining p_i^* to be the highest value among p_{ijk} corresponding to the highest probability to minimize a cost function related to the process under consideration

$$P^* = \{p_i^* = p_{ijk}, x_i \xrightarrow{p_i^*} y_{jk}/p_{ijk} = \max_{q,r} p_{iqr},\ i = 1,..n\} \tag{8}$$

where $x_i \varepsilon L(G_i)$ is an input string and $y_{jk} \varepsilon L(G_{oj})$ is an output string. Sequential updating of the subjective probabilities p_{ijk}, may be interpreted as learning when new performance information is available. It can be obtained through two stochastic approximation algorithms:

$$p_{ijk}(t+1) = p_{ijk}(t) = \gamma_{ijk}(t+1)\ [\xi_{ijk}(t) - p_{ijk}(t)]$$

$$J_{ijk}(n_{ijk}+1) = J_{ijk}(n_{ijk}) + \beta_{ijk}(n_{ijk}+1)\ [C(n_{ijk}+1) - J_{ijk}(n_{ijk})] \tag{9}$$

$$\xi_{ijk}(t) = \begin{cases} 1 & \text{if } J_{ijk} = \min\limits_{q,m} J_{iqm} \\ 0 & \text{otherwise} \end{cases}$$

where J_{ijk} is the performance estimate, C is the observed cost, n_{ijk} is the number of occurence of the event (x_i, y_{jk}) and $\gamma_{ijk}(t+1)$ and $\beta_{ijk}(n_{ijk}+1)$ are sequences satisfying Dvoketcky's condition for convergence of the algorithms. Given the ith input (command) the entropy associated with the output of the schema is:

$$H_i = \sum_j \sum_k H_{ijk} = \sum_j \sum_k p_{ijk} \ln p_{ijk} \tag{10}$$

Several special decision schemata and associated decision codes have been proposed by Saridis and Graham (1984). However, the easiest to implement is the so-called Vocabulary Optimal Decision Schema in which the input and output languages retain the same syntax, e.g., structural form, but there are several output terminal vocabularies to be selected for different tasks. They can all be implemented with a transducer automator and require only short term memory to implement the function of learning (see Figure 2). The coordination level is essential for dispatching organizational information to the next level, the execution level.

6. THE THEORY OF INTELLIGENT MACHINES-THE CONTROL LEVEL

This and the next section are dedicated to show that the cost of control problem at the hardware level can be expressed as an entropy which measures the uncertainty of selecting an appropriate control to execute a task. By selecting an optimal control, one minimizes the entropy e.g., the uncertainty of execution. The entropy may be viewed in the respect as an energy in the original sense of Botlzmann, as in Saridis (1984).

In order to utilize entropy as a measure of the performance (energy) of a control system, a probabilistic interpretation must be given to the problem.

The similarities of the Theories of Thermodynamics and Dynamical Systems, suggest a parallel to be drawn. As in the case of Thermodynamics, where the average Lagrangian is related to an entropy, the performance measure of a control system usually described by a generalized energy function should be expressed as an entropy, and thus integrate it with the rest of the measures of a hierarchically intelligent machine. The claim is proved in the next section. Optimal control theory, Athans and Falb (1965) has utilized a non-negative functional of the states of the system $x(t) \in \Omega_x$ the state space, and a specific control $u(x,t) \in \Omega_u$ the set of all admissible controls, to define the performance measure for some initial conditions $(x(t),t)$, representing a generalized energy function, of the form,

$$V(x,t) = \int_t^{t_f} L(x,u(x,\tau),\tau)\, d\tau \tag{11}$$

where $L(x,u(x,\tau),\tau) > 0$, subject to differential constraints dictated by the underlying process

$$\dot{x} = f(x,u(x,t),t) \quad x(t_0) = x_s, \; x(t_f) \; \varepsilon M_f \tag{12}$$

with M_f a manifold in Ω_x. The trajectories of the system (12) are defined for a fixed but arbitrarily selected control u(x,t) from the set of admissible controls Ω_u.

In order to express the control problem in terms of an entropy function, one may assume that the performance measure V(x,u(x,t),t) is distributed in Ω_x according to the controls u(x,t) and the initial conditions (x,t). The optimal performance corresponds to the maximum point of the associated density function p(x,u(x,t),t).

Then, an entropy function H_c, corresponding to the above density, may be defined to represent the performance of the system,

$$H_c(u) = - \int_{\Omega_x} p(x,u(x,t),t) \ln p(x,u(x,t),t) \, dx \tag{13}$$

Using the marginal probabilities x to include u as a random variable and using the chain rule, H_c becomes

$$H(u)_c = - \int_{\Omega_x} \int_{\Omega_u} p(x/u) \, p(u) \ln[p(x/u)p(u)] dx \, du \tag{14}$$

Considering that the initial conditions are independent of u p(x/u) = p(x) and

$$\begin{aligned} H_c(u) &= - \int_{\Omega_x} \int_{\Omega_u} p(x)p(u) \, [\ln p(x) + \ln p(u)] \, dx \, du = \\ &= - \int_{\Omega_x} p(x)dx \int_{\Omega_u} p(u)du \, [\ln p(x) + \ln p(u)] = \\ &= - \int_{\Omega_x} p(x) \ln p(x)dz - \int_{\Omega_x} p(u)\ln p(u) \, du = H_x + H_u \end{aligned} \tag{15}$$

Here H_u is the entropy corresponding to the subjective probability of selecting the control u(x,t) and H_x is the entropy corresponding to the objective probabilities defined by the uncertainty of the initial conditions x,t, of the system. When initial conditions are given e.g., p(x,t) = 1, lnp(x,t) = 0, the entropy $H_x = 0$ and

$$H_c(u) = H_u = - \int p(u) \ln p(u)\, du \tag{16}$$

This result shows that H_c is the entropy of selecting the control u(x,t) from the set of admissible controls Ω_u. Due to (13), minimizing H_c is then equivalent to selecting $u^*(x,t)$, the optimal control. What is left is to show that $H_c(u)$ of equation (15) is equivalent to the performance measure of equation (11). This is done in the next sections.

7. THE MEASUREMENT OF A CONTROL SYSTEM'S PERFORMANCE WITH ENTROPY

The formulation and derivations of the previous section suggested a probabilistic interpretation of the control problem: One may assume that the values of the performance measure V(x,t) are distributed according to the assignment of the admissible control $u(x,t) \in \Omega_u$. Influenced by the discussion of Section 3, the probability density p(x,u(x,t),t) associated with the distribution of the controls may be selected to be an exponential, and to satisfy the incompressibility condition for densities eq. (3). It remains to show that the entropy H_c, associated with the density p(x,u(x,t),t), represents the average performance measure of the control problem V(x,t) associated with the distribution of u(x,t) over Ω_u or that H_c is equivalent to the entropy of selecting the control u(x,t) from Ω_u. This is done as follows: Define a probability density function (p(x,u (x,t),t) corresponding to an assigned control u(x,t) by

$$p(x,u(x,t),t) = Ce^{-V(x,u(x,t),t)};$$

$$\int_{\Omega_x} p(x,u(x,t),t)\, dx = 1 \tag{17}$$

where C is a normalizing factor.

This probability density p(x,u(x,t),t) must satisfy the same incompressibility condition (3) in the state space, as in the case of statistical thermodynamics and information theory. This condition, applied to the above density, yields a Liouville type equation,

$$\frac{\partial p}{\partial t} + \frac{\partial p}{\partial x}\dot{x} + L(x,u(x,t),t) = 0 \tag{18}$$

where

$$\frac{\partial p}{\partial t} = - \frac{\partial V}{\partial t} p, \quad \frac{\partial p}{\partial x} = - \frac{\partial V}{\partial x} p, \tag{19}$$

Substituting back into (16) yields

$$\left\{ \frac{\partial V}{\partial t} + \frac{\partial V}{\partial x}^{T} f(x,u(x,t),t) + L(x,u(x,t),t) \right\} p(x,u(x,t)) = 0 \tag{20}$$

Since the probability density function is an exponential, $p(x,u(x,t)) \neq 0$ anywhere in Ω_x. Therefore, the first term (20) must be equal to zero. This implies that the incompressibility condition is equivalent to the generalized Hamilton-Jacobi equation of the control problem, obtained by Saridis and Lee (1979).

$$\begin{aligned} &\frac{\partial V}{\partial t} + \frac{\partial V}{\partial x}^{T} f(x,u(x,t),t) + L(x,u(x,t),t) = 0 \\ &V(x(t_f),t_f) = 0 \end{aligned} \tag{21}$$

Based on these equations the optimal control problem may be stated as follows. The optimal control $u^*(x,t)$, when it exists, and its performance criterion $V^*(x,t)$ satisfy equation (21), in its minimum form:

$$\begin{aligned} &\frac{\partial V}{\partial t} + \frac{\partial V}{\partial x}^{T} f(x,u^*(x,t),t) + L(x,u^*(x,t),t) = \\ &\frac{\partial V}{\partial t} + \min_u \left[\frac{\partial V}{\partial x}^{T} f(x,u,t) + L(x,u,t) \right] = 0 \end{aligned} \tag{22}$$

Having the incompressibility condition satisfied, the density function $p(x,u(x,t),t)$ selected for the control problem, may be substituted in the expression of the entropy H_c in eq. (13) representing the entropy for the control problem for a particular control $u(x,t)$. This yields

$$H_c = \int_{\Omega_x} p(x,u(x,t),t)\, V(x,u(x,t),t)\, dx = E_x \{V(x,u(x,t),t)\} \tag{23}$$

The result obtained indicates that the average, with respect to initial state performance measure of a feedback control problem corresponding to a specifically selected control, can be expressed as an entropy function.

Furthermore, the optimal control u^* that minimizes $V(x,u(x,t),t)$, maximizes $p(x,u(x,t),t)$, and consequently minimizes the average performance measure.

$$u^*: E_x \{V(x,u^*(x,t),t)\} = \underset{u}{\text{Max}} \left[-\int_{\Omega_x} V(x,u(x,t),t)\ p(x,u(x,t),t)\ dx\right] =$$

$$= \underset{u}{\text{Min}} \int_{\Omega_x} V(x,u(x,t),t)\ p(x,u(x,t),t)dx \qquad (24)$$

This statement establishes equivalent measures between information theoretic and optimal control problems and provides the information and feedback control theories with a common measure of performance. Entropy satisfies the additive property and any system composed of a combination of such subsystems will be optimal by minimizing its total entropy. Therefore the design of an intelligent machine may be formulated as the mathematical programming problem, of minimizing the sum of the entropies of all the subprocesses involved subject to the constraints inposed by the "intelligent machine."

8. THE COMPLETE INTELLIGENT MACHINE: A CASE STUDY IN ROBOTS

The previous sections discussed the details of the generic structures of a hierarchically intelligent machine and the analytic methodologies involved in controlling them to perform a series of tasks with minimum interaction with a human operator. However, the complete structure of such a machine, cannot be defined in general, but must be specified for each particular case under consideration.

Even at the present time there is a large variety of applications for intelligent machines. Automated material handling and assembly in an automated factory, automated inspection, sentries in a nuclear containment are some of the areas where intelligent machines have and will find a great use. One of the most important applications though is the unmanned space exploration where because of the distances involved, autonomous anthropomorphic tasks must be executed and only general commands and reports of executions may be communicated.

Such tasks are suitable for "intelligent robots" a type of intelligent machines capable of executing anthropomorphic tasks in unstructured uncertain environments. They are structured usually in a human like shape and are equipped with vision and other tactile sensors to sense the environment, two areas to execute tasks and locomotion for appropriate mobility in the unstructured environment. The controls of such a machine are performed according to the theory of Intelligent Machines previously discussed, Saridis and Stephanou (1977), Saridis (1983). The three levels of controls, obeying the Principle of Decreasing Intelligence and Increasing Precision, are implemented with appropriately selected feedback, as shown in Figure 3.

The organization level, the level of highest intelligence, receives commands from headquarters to execute a series of tasks in an unstructured environment. It contains one or more planning units, sensory data processors and a long term memory for knowledge flow management and organization of the tasks required to execute the command.

The outputs of the organization level are channels to the coordination level composed of several decision schemata and short term memories (buffers) each for every hardware mechanism of the next level. Learning and performance selection are endowed in each module so that adaptation and coordination is possible for new unexpected situations. Selective feedback from the sensors is available to this level in order to make its operation autonomous from the organization when required.

The hardware execution level is composed of controllers that drive the mechanical areas, locomotion, and the various sensors that provide feedback information from the environment. Their performance is coordinated by the previous level by selecting performance criteria with the appropriate specifications, constraints for object avoidance and terminal manifolds for task accomplishments. Adaptive control is also possible for uncertain handling situations.

The overall structure thus modeled, may be optimized by minimizing the total entropy, occurred hierarchically as one moves backwards on the decision tree of the three levels of the system, seen in Figure 4. The individual entropies of each device are obtained from the procedures discussed in section 4, 5 and 6 of this work. The entropy accrued at each level, represents the imprecision (uncertainty) and increases as one moves up in the hierarchy, in accordance with the principle. On the other hand, the flow of knowledge at the highest level is maximum in order to handle the total entropy again in accordance with the principle. Its function is to find the proper polls on the decision tree which minimizes the total entropy. This process resembles somehow the process of biological systems, deiscussed by Prigogine (1980).

Simulations and experimental implementation are presently in progress at the Robotics and Automation Lab at RPI for verification and evaluation of the theory.

9. CONCLUSIONS

The problem of controlling systems in uncertain environments with minimum interaction with a human operator has been presented, as the problem of designing an "intelligent machine". A methodology has been developed to formulate the "intelligent machine" as a mathematical programming problem as using the aggregated entropy of the system as its performance measure. The higher levels of the machine structured according to the Principle of Increasing Precision with Decreasing Intelligence can adopt performance measures easily expressed as entropies. This work is based to a major extent on the equivalence between the control performance and the entropy of selecting an appropriate control, thus integrating the execution level of the machine into the overall mathematical programming problem. Optimal solutions of the "intelligent machine" problem can be obtained by minimizing the overall entropy of the system. The entropy formulation presents a tree-like structure for this decision problem very appealing for real-time computational solutions. Reasonable approximations may thus be obtained using one step ahead entropies only when the system complexity increases.

REFERENCES

Albus, J. S., "A New Approach to Manipulation Control: The Cerebellar Model Articulation Controller", Transactions of ASME, Journal of Dynamic Systems, Measurement and Control, Vol. 97, pp. 220-227, September (1975).

Ashby, W. R., An Introduction to Cybernetics, J. Wiley & Sons, Science Edition, New York, (1965).

Athans, M., Falb, P., Optimal Control, McGraw Hill, New York, (1965).

Bayes, T., "An Essay Toward Solving a Problem in the Doctrine of Chances", Philosophical Transactions of the Royal Society, 53, pp. 370-418 (1763).

Boltzmann, L., Populari Schriften, Leipzig (1905).

Conant, R. C., "Laws of Information Which Govern Systems", IEEE Transactions on SMC, Vol. SMC-6, No. 4, pp. 240-255, April (1976).

Findeisen, Bailey, et. al., Control and Coordination in Hierarchical Systems, John Wiley & Sons, New York, (1980).

Hayes-Roth, et. al., Building Expert Systems, Addison-Wesley, New York, (1983).

Heisenberg, W., Z. Physik 33, p. 879 (1925).

Prigogine, From Being to Becoming, W. H. Freeman and Co., San Francisco, (1980).

Saridis, G. N., Stephanou, H. E., "A Hierarchical Approach to the Control of a Prosthetic Arm", IEEE Transactions on SMC, Vol. SMC-7, No. 6, pp-407-420 June (1977).

Saridis, G. N., "Toward the Realization of Intelligent Controls", Proceedings of IEEE, Vol. 67, August (1979).

Saridis, G. N., Lee, C. S. G., "Approximation Theory of Optimal Control for Trainable Manipulators", IEEE Transactions on SMC, Vol. SMC-8, March (1979).

Saridis, G. N., Blumberg, J., "Entropy As A Cost Criterion for Hierarchical Intelligent Control", RPI/RAL Report 17, August (1983).

Saridis, G. N., "Intelligent Robotic Control", IEEE Transactions on AC, Vol. AC-28, No. 5, pp. 547-557, May (1983).

Saridis, G. N., Graham, J. H., "Linguistic Decision Schemata for Intelligent Robots", Automatica, Vol. 20, No. 1, pp. 121-126 (1984).

Saridis, G. N., "Control Performance as an Entropy: An Integrated Theory of Intelligent Machines", Proceedings of International Conference on Robotics, Atlanta, Georgia, March (1984).

Shafer, G., A Mathematical Theory of Evidence, Princeton University Press, Princeton, N.J. (1976).

Shannon, C., Weaver, W., The Mathematical Theory of Communication, Illini Books, (1963).

Stephanou, H. E., Lu, S. Y., "Measuring Concensus Effectiveness by a Generalized Entropy Criterion", Proc. Conference on Artificial Intelligence, Denver, Co. (1984).

Whinston, P., Artificial Intelligence, Addison Wesley, Reading, MA, (1977).

Wiener, N., Cybernetics, MIT Press, Cambridge, MA, (1962).

ACKNOWLEDGEMENT

This work was supported by NSF Grant ENG ECS - 8312179.

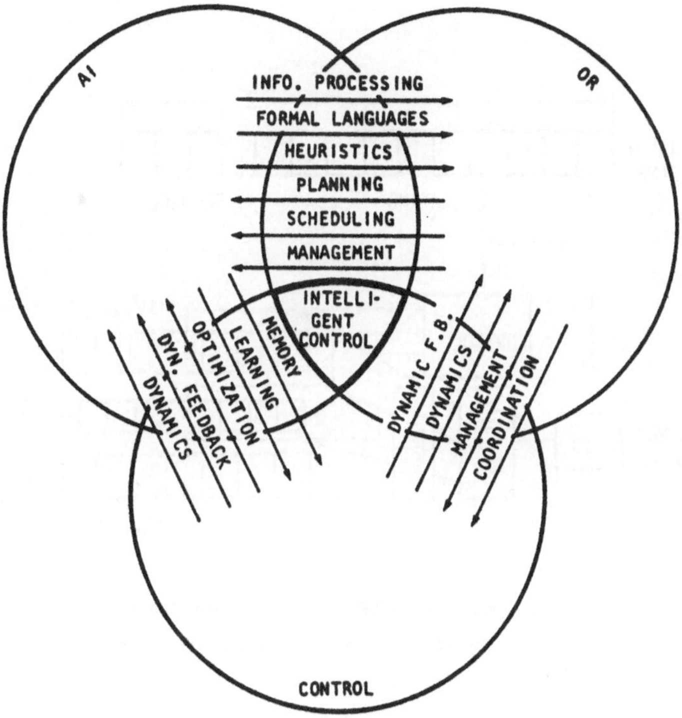

FIGURE 1. INTERACTION OF ARTIFICIAL INTELLIGENCE, OPERATIONS RESEARCH, AND CONTROL THEORY, AND THE RESULTING INTELLIGENT CONTROL.

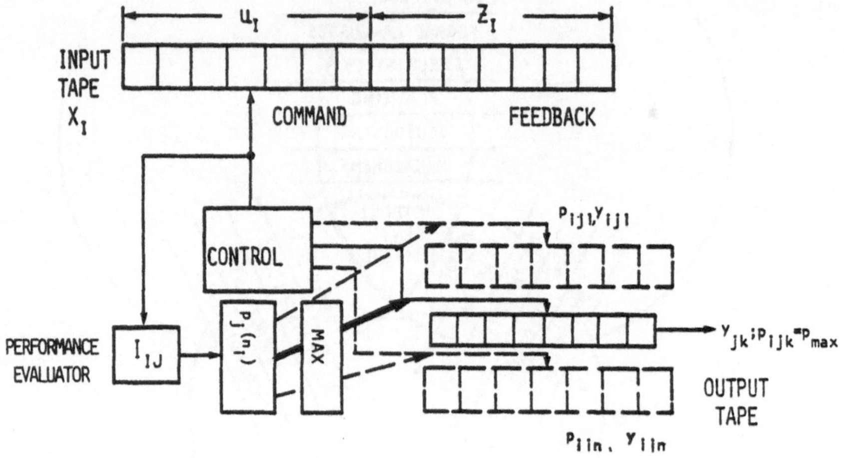

FIGURE 2. SCHEMATIC REPRESENTATION OF A TRANSDUCER AUTOMATON FOR THE REALIZATION OF A "DECISION CODE".

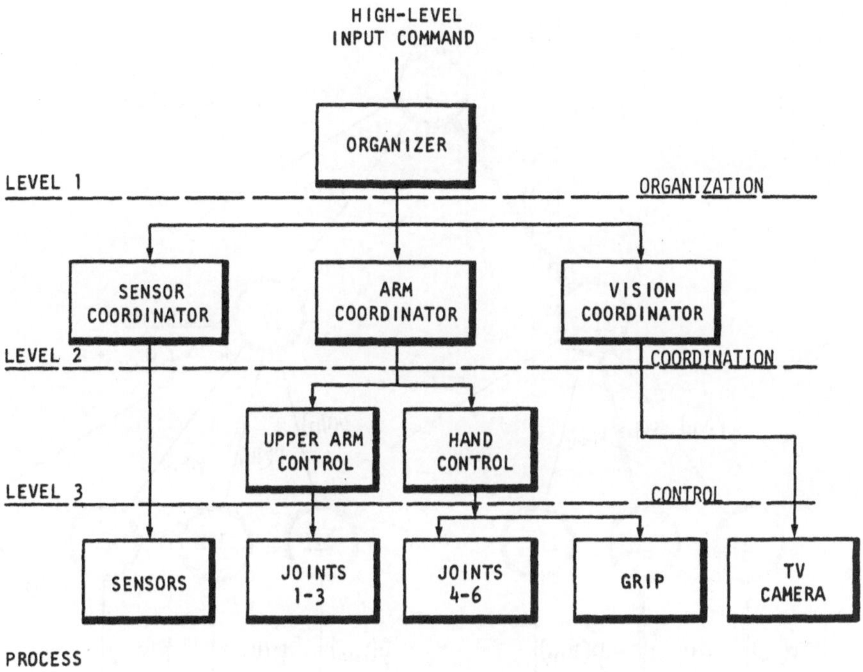

FIGURE 3. HIERARCHICALLY INTELLIGENT CONTROL FOR Q MANIPULATOR WITH VISUAL FEEDBACK

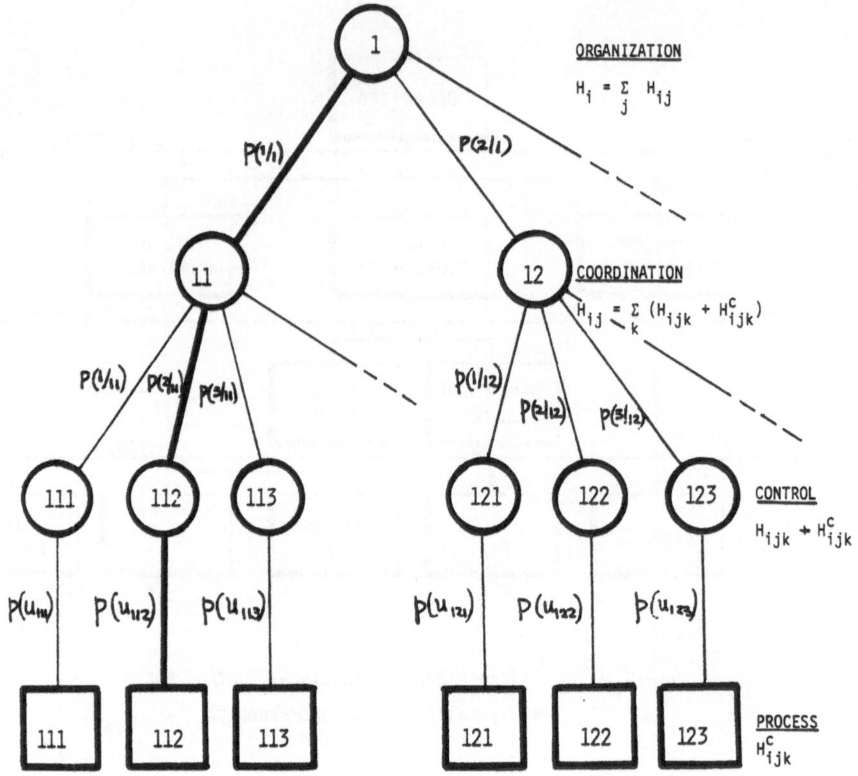

FIGURE 4. DECISION TREE FOR A HIERARCHICALLY INTELLIGENT CONTROL WITH ENTROPIES AT THE NODES

Lecture Notes in Control and Information Sciences

Edited by M. Thoma

Vol. 43: Stochastic Differential Systems
Proceedings of the 2nd Bad Honnef Conference
of the SFB 72 of the DFG at the University of Bonn
June 28 – July 2, 1982
Edited by M. Kohlmann and N. Christopeit
XII, 377 pages. 1982.

Vol. 44: Analysis and Optimization of Systems
Proceedings of the Fifth International
Conference on Analysis and Optimization of Systems
Versailles. December 14–17, 1982
Edited by A. Bensoussan and J. L. Lions
XV, 987 pages, 1982

Vol. 45: M. Arató
Linear Stochastic Systems
with Constant Coefficients
A Statistical Approach
IX, 309 pages. 1982

Vol. 46: Time-Scale Modeling of Dynamic Networks
with Applications to Power Systems
Edited by J. H. Chow
X, 218 pages. 1982

Vol. 47: P. A. Ioannou, P. V. Kokotovic
Adaptive Systems with Reduced Models
V, 162 pages. 1983

Vol. 48: Yaakov Yavin
Feedback Strategies for Partially
Observable Stochastic Systems
VI, 233 pages, 1983

Vol. 49: Theory and Application of Random Fields
Proceedings of the IFIP-WG 7/1
Working Conference
held under the joint auspices of the
Indian Statistical Institute
Bangalore, India, January 1982
Edited by G. Kallianpur
VI. 290 pages. 1983

Vol. 50: M. Papageorgiou
Applications of Automatic Control Concepts
to Traffic Flow Modeling and Control
IX, 186 pages. 1983

Vol. 51: Z. Nahorski, H.F. Ravn, R.V.V. Vidal
Optimization of Discrete Time Systems
The Upper Boundary Approach
V, 137 pages 1983

Vol. 52: A. L. Dontchev
Perturbations, Approximations and Sensitivity Analysis
of Optimal Control Systems
IV, 158 pages. 1983

Vol. 53: Liu Chen Hui
General Decoupling Theory of Multivariable
Process Control Systems
XI, 474 pages. 1983

Vol. 54: Control Theory for Distributed
Parameter Systems and Applications
Edited by F. Kappel, K. Kunisch,
W. Schappacher
VII, 245 pages. 1983.

Vol. 55: Ganti Prasada Rao
Piecewise Constant Orthogonal Functions
and Their Application to Systems and Control
VII, 254 pages. 1983.

Vol. 56: Dines Chandra Saha, Ganti Prasada Rao
Identification of Continuous
Dynamical Systems
The Poisson Moment Functional
(PMF) Approach
IX, 158 pages. 1983.

Vol. 57: T. Söderström, P. G. Stoica
Instrumental Variable Methods
for System Identification
VII, 243 pages. 1983.

Vol. 58: Mathematical Theory of
Networks and Systems
Proceedings of the MTNS-83 International
Symposium
Beer Sheva, Israel, June 20–24, 1983
Edited by P. A. Fuhrmann
X, 906 pages. 1984

Vol. 59: System Modelling and Optimization
Proceedings of the 11th IFIP Conference
Copenhagen, Denmark, July 25-29, 1983
Edited by P. Thoft-Christensen
IX, 892 pages. 1984

Vol. 60: Modelling and Performance
Evaluation Methodology
Proceedings of the International Seminar
Paris, France, January 24–26, 1983
Edited by F. Bacelli and G. Fayolle
VII, 655 pages. 1984

Vol. 61: Filtering and Control of Random
Processes
Proceedings of the E.N.S.T.-C.N.E.T. Colloquium
Paris, France, February 23–24, 1983
Edited by H. Korezlioglu, G. Mazziotto, and
J. Szpirglas
V, 325 pages. 1984

Lecture Notes in Control and Information Sciences